符号中国 SIGNS OF CHINA

旗　袍

CHI-PAO

"符号中国"编写组 ◎ 编著

中央民族大学出版社
China Minzu University Press

图书在版编目(CIP)数据

旗袍：汉文、英文 /"符号中国"编写组编著. —北京：中央民族大学出版社，2024.8
（符号中国）
ISBN 978-7-5660-2325-4

Ⅰ.①旗…　Ⅱ.①符…　Ⅲ.①旗袍-介绍-中国-汉、英　Ⅳ.①TS941.717

中国国家版本馆CIP数据核字（2024）第016879号

符号中国：旗袍　CHI-PAO

编　　著	"符号中国"编写组
策划编辑	沙　平
责任编辑	陈　琳
英文指导	李瑞清
英文编辑	邱　械
美术编辑	曹　娜　郑亚超　洪　涛
出版发行	中央民族大学出版社
	北京市海淀区中关村南大街27号　邮编：100081
	电话：（010）68472815（发行部）　传真：（010）68933757（发行部）
	（010）68932218（总编室）　　　　（010）68932447（办公室）
经销者	全国各地新华书店
印刷厂	北京兴星伟业印刷有限公司
开　　本	787 mm×1092 mm　1/16　印张：10.75
字　　数	149千字
版　　次	2024年8月第1版　2024年8月第1次印刷
书　　号	ISBN 978-7-5660-2325-4
定　　价	58.00元

版权所有　侵权必究

"符号中国"丛书编委会

唐兰东　巴哈提　杨国华　孟靖朝　赵秀琴

本册编写者

徐　冬

前言 Preface

旗袍是中国传统女性服饰的代表，是最能体现中国女性魅力的服装之一。旗袍有着含蓄的东方美，能够表现出女性的独特韵味，线条简洁，色彩灵动，风格优雅，细腻的面料贴身而适体。可以巧妙地勾勒出女性身材优美的曲线，使女人优雅、柔美、贤淑、妩媚的风情和气质尽显无遗。

旗袍从清代发展到今天，已经有三百多年的历史。现在，它更以崭新的面貌出现在世界时装展台，展示着蓬勃

Chi-pao, with high neck and slit skirt, is a representative of traditional Chinese dresses for women. It is one of the costumes that can best accentuate the charm of Chinese women. Chi-pao reflects the reserved oriental beauty and can demonstrate the unique grace and elegance of women. It has simple lines, flexible colors, graceful styles and refined fabrics, and can set off women's figure so finely and displaying women's grace, beauty, virtue and charm so completely.

Since its advent in

的生命力和迷人的魅力。

　　本书将对旗袍的独特魅力，包括旗袍的演变历史、精湛的制作工艺、多变的配饰，以及穿着旗袍时应该注意的事项进行全面的展示。通过本书，您可以领略并亲自演绎旗袍之美。

the Qing Dynasty (1616–1911), Chi-pao has been developing for over three hundred years. Now, in a brand-new appearance, it shines on all fashion stages in the world, fully demonstrating its strong vitality and charming glamour.

　　This book will showcase the unique charm of Chi-pao by touching upon all aspects related to it, including its history, exquisite techniques for its making, diversified adornments and tips for its wearers. Through this book, readers can experience and personally interpret the beauty of Chi-pao.

目 录 Contents

旗袍的由来
Origin of Chi-pao 001

历史上的袍服
Gown-style Dresses in History 002

清代的旗袍
The Chi-pao in the Qing Dynasty 004

旗袍的演变
Evolution of Chi-pao 013

旗袍之美
Beauty of Chi-pao 037

形之美
Beauty of the Shape 038

工艺之美
Beauty of the Make 052

质之美
Beauty of the Texture 079

选购适合自己的旗袍
Choosing a Chi-pao Right for You 093

旗袍的选购
Selection and Purchase of a Chi-pao 094

旗袍的定制
Tailor-made Chi-pao 115

旗袍的配饰
Adornments for Chi-pao 122

旗袍的保养
Taking Good Care of Your Chi-pao 141

附录：现代旗袍鉴赏
Appendix: Appreciation of Modern Chi-pao ... 145

旗袍的由来
Origin of Chi-pao

　　中国素有"衣冠王国"之美誉，中国人以自己的聪明才智创造出无数具有中国特色的服饰。中国民族众多，不同地域的民俗风情、人们的衣着习惯各具特色，因此可以说，中国的服饰文化是多民族服饰融合而形成的。

China has long been referred to as the Kingdom of Dresses. The Chinese people have created countless dresses and adornments with Chinese characteristics using their wisdom and talents. China has many ethnic groups, and different places have their folk customs, including those in clothing. It is safe to say that the clothing culture of China is formed by the merging of the costumes of all the ethnic groups in the country.

> 历史上的袍服

旗袍由清代满族妇女所穿的袍服演变而来。袍服是中国古代的传统服饰之一，有着悠久的历史，早在先秦时期（公元前221年以前）就已经出现了。

袍服作为正式的服装始于东汉，汉代定袍服为礼服。唐太宗

> Gown-style Dresses in History

Chi-pao has evolved from the gown-style dress (gown sounds like *pao* in Chinese) worn by Manchu women in the Qing Dynasty (1616–1911). In fact, the gown-style dress is one of the traditional Chinese dresses. It appeared even before the Qin Dynasty (221 B.C.-206 B.C.).

The gown-style dress had been recognized as formal attire since the

- 明代杜堇的《宫中图》（局部）
  画卷描绘了明朝时宫中嫔妃们的日常生活：宫廷画师正在画像，周围五人或坐或立，神态各异。侍女们身着直身宽筒形袍服。
  Inside the Palace (Partial) by Du Jin in the Ming Dynasty (1368-1644)
  The painting depicts a daily-life scene of the imperial concubines inside the palace of the Ming Dynasty (1368-1644). An imperial painter is drawing a portrait. The five female figures around him have different expressions on their faces but they all wear straight loose cylindrical gown-style dresses.

在位时（627—649），曾诏定全国除元旦、冬至的朝会及祭祀，一律着袍服。随着袍服之制的普及，袍服逐渐成为一种稳定的传统服装样式，并为历代所沿用。

历代袍服在形制上有所变化。早期中原地区通用的袍服一般衣身宽松，衣长至脚踝，袖肥阔，袖口处呈收缩状，臂肘处形呈圆弧状，大袖翩翩，造型美观；而流行于北方少数民族地区的袍服与中原袍服恰好相反，一般为圆领、窄袖，紧身而合体。

随着中原地区与少数民族地区的文化交融，少数民族的袍服传入中原，为汉族所接受，并发展成大众服饰，旗袍就是其中的一种。

- **宋代宫廷画**

此图中，右侧为穿长裙的贵妇，左侧为穿袍服的侍女。

A Court Painting of the Song Dynasty (960-1279)

In the picture, the noble woman on the right wears a full-length skirt while the maid on the left wear gown-style dresses.

Eastern Han Dynasty (25–220). In the Tang Dynasty (618-907), Emperor Taizong (on the throne from 627 to 649) issued an imperial edict to the effect that all people in the country should wear the gown-style dress except on the New Year's Day, court meeting on the Winter Solstice, and sacrifice-offering ceremonies. With its popularization, the gown-style dress gradually developed into a stable traditional costume that was accepted from dynasty to dynasty.

The gown-style dress had varied forms in different dynasties. In the early period, the gown-style dress in the Central Plains had a spacious design. Its hem dropped to the wearer's ankles and its loose and spacious sleeves had contracted cuffs and curved elbows. Its design was indeed beautiful. Totally different from it was the gown-style dress popular among northern ethnic groups, which had a round collar, narrow sleeves, and a close-fitting design.

With cultural communication between the Central Plains and the areas inhabited by ethnic groups, the gown-style dresses of the ethnic groups were spread to the Central Plains and accepted by the Han people. These dresses then developed into popular clothing and Chi-pao was one of them.

> 清代的旗袍

旗袍指满族人所着之袍。1616年，满族首领努尔哈赤建立了后金政权，推行八旗制度（清代满族的社会组织形式，将满族人编制在各

- 清代初期的女性画像（叶衍兰绘）
  A Female Portrait in the Early Period of the Qing Dynasty (by Ye Yanlan)

> The Chi-pao in the Qing Dynasty

Chi-pao derived its name from the fact that it was worn by the Manchu people. In 1616, Nurhachi, head of the Manchu people, established the Later Jin Dynasty and adopted the Eight Banners System (a social organization form of the Manchu people in the Qing Dynasty (1616-1911), and according to it, the Manchu people were categorized into eight groups, or eight banners, namely Yellow, White, Red, Blue, Bordered Yellow, Bordered White, Bordered Red and Bordered Blue) The clothes worn by the Manchu people were called Chi-clothing.

Before 1644, the Manchu people lived in the Changbai Mountain and the Songhua River Valley in the northeast of China. They made a living by fishing and hunting. The earliest Chi-pao had a simple style and structure, with one

旗内，包括正黄、正白、正红、正蓝、镶黄、镶白、镶红、镶蓝），通称"八旗"，满族人所穿的服装统称为"旗装"。

在1644年入关以前，满族人长期居住在中国东北部的长白山、松花江流域，以游牧、渔猎为生。因此，早期的旗袍样式和结构都比较简单，不分上下，宽身、直筒，与汉族服装上衣下裳的结构有明显的区别。旗袍的基本款式特点为圆领、大襟、窄袖，带纽扣和腰带。袖口窄小是此时的旗袍最明显的特征，袖头形状与马蹄类似，又称"马蹄袖"，也称"箭袖"。冬季或作战时，袖子能罩住手背，既保暖，又不影响拉弓射箭，平时绾起来，还能作为装饰。此外，旗袍四面开衩的造型便于骑马；束腰带既可以起御寒作用，还可以在前襟内存放干粮。

1644年，满族入关，定都北京，建立了清王朝。随着政权的初步稳固，统治者开始制定并完善清朝的衣冠制度，强制推行服制改革，规定除了汉族女子居家时不必穿着旗装，其余场合，所有人一律着旗装。

straight and cylindrical piece covering the whole body. It was apparently different from the two-piece design of the Han clothing. A typical Chi-pao at that time had a round collar, a large front piece,

- 《孝庄文皇后常服像》（清）
清代满族女性的常服袍是旗袍的前身。这幅画像反映了清代初期旗袍的基本款式：圆领、大襟、左右开衩，衣身宽松，衣摆掩足。边饰十分简洁，只表现为衣摆处有黑色镶边。

A Portrait of Empress Dowager Xiaozhuang in an Informal Dress (Qing Dynasty, 1616-1911)

The informal gown for Manchu women in the Qing Dynasty was the predecessor of Chi-pao. This portrait reveals some basic features of the Chi-pao in the early period of the Qing Dynasty: round collar, large front piece, and slits at both sides. The Chi-pao at that time was loose and its hem covered the wearer's feet. Its edging was simple, with only a black lace along the hem.

• 清代身穿各式旗袍的官廷贵妇

清朝末年，旗袍的风格由简洁变得繁复，讲究装饰，衣襟、袖口、衣摆等部位还以汉族的刺绣工艺和吉祥纹样进行装饰，颜色、品种、图案都十分丰富。

Noble Women in Various Styles of Chi-pao in the Qing Dynasty (1616-1911)

In the final years of the Qing Dynasty, the design of Chi-pao turned from simple to complex and laid more stress on decoration. The embroidery techniques and various auspicious patterns of the Han people were adopted. These new decorations were on the front piece, cuffs, and hem of the Chi-pao.

满族人进入中原后，生活方式发生了很大的变化，由游牧、狩猎变为稳定的农耕，再加上满、汉文化的不断交流和相互影响，使得满族的服饰也悄然发生着变化。

旗袍的袖口由紧而窄逐渐转为宽松，被称为"倒大袖"；四面开衩变为两面开衩或不开衩；圆领口或较低的立领变为较高的立领；在用料上，从以前的厚重而耐磨的毛

narrow sleeves, several buttons and a waist belt. The narrow cuff was the most obvious feature. The cuff was shaped like a horse hoof, hence its two other names Horse Hoof Cuff and Arrow Cuff. In winter, the sleeves could cover the back of the wearer's hands, keeping the warmth but not hampering arrow shooting. In normal times, they could be rolled up as a decoration. In addition, the Chi-pao had slits on four sides, a feature good for horse riding. Its waist belt could not only help preserve warmth, but allow storage of food in the front piece.

In 1644, Beijing was selected as the capital of the Qing Dynasty. After their regime was stabilized, the Qing rulers began stipulating and improving the clothing institution of the Qing Dynasty by forcing a clothing reform on the people. They ordered that all people should wear Chi-clothing, with only one exception: Han women did not have to wear Chi-clothing in their homes.

After entering the Central Plains, the Manchu people experienced a huge change in their lifestyle. They turned

皮变为以轻便的棉布、绸缎为主；在用色上，由起初的色调自然而淡雅发展到后来逐渐讲求花色搭配，色彩光鲜而亮丽；在装饰上，由纹饰简洁发展到后来在衣领、衣袖、衣襟等部位镶以、滚以花边。

from a nomadic and hunting lifestyle to a stable farming lifestyle. In addition, the nonstop cultural communication and mutual influence between the Han and Manchu peoples also contributed to the change of Manchu clothing.

The cuffs of Chi-pao have gradually turned from tight to loose and were known as "reversed large sleeves". The four-side slits became two-side slits and some Chi-pao had no slit at all. The low round collar became higher. The previous heavy and wear-resisting fur was replaced by lighter cotton cloth and silk. The color of Chi-pao also changed from simple to diversified and bright. At first, there were simple pattern adornments. Later, the lace trimmings and embroidered borders were seen on the collar, cuff and front pieces of the clothes.

• 清代汉族女性的裙套装
A Dress for Han Women in the Qing Dynasty (1616-1911)

- **清代慈禧太后的画像**

  画面中的慈禧一身珠光宝气，穿着一件杏黄色"寿"字纹绸缎旗袍，工艺精细，装饰华丽，上面点缀着数颗又大又圆的珍珠。

  ### A Portrait of the Empress Dowager Cixi in the Qing Dynasty

  In the portrait, Cixi is adorned with brilliant jewels and pearls. She wears an apricot yellow silk Chi-pao with the patterns formed by the Chinese character for longevity. The Chi-pao shows refined tailoring and gorgeous adornment. It is mounted with several big and round pearls.

- **清代中后期汉族女子的日常装扮**

  画面中的汉族女子着装为上袄下裙式，发髻低矮，服饰色彩亮丽，装饰华美。上身为圆领、大襟、宽袖袍服，下身为裙。袄的衣缘带有大镶边，两侧有明显的如意云头纹。衣摆过膝，体现了满汉融合的趋势。图中的女子穿一双尖而窄的鞋子，形如弯弓。

  ### Daily Attire of Han Women in the Middle and Late Periods of the Qing Dynasty

  The Han woman in the picture wears a jacket and a skirt. Her hair is made into a low bun and her clothes have bright colors and gorgeous decorations. Her jacket has a round collar, a large front piece, and loose sleeves. The jacket has a large laced hem with obvious auspicious cloud patterns on both sides. The jacket hem drops far below her knees, demonstrating the trend towards the integration of the Han and Manchu styles. On her feet is a pair of narrow, tapered, and bow-shaped shoes.

领：早期的清代旗袍多为圆领。
Collar: Most Chi-pao in the early period of the Qing Dynasty had a round collar.

袖：衣袖短、宽且直，与衣身相接。此时的旗袍均为连袖。
Sleeves: Short, wide and straight, the sleeves and the main part of the Chi-pao form an integral whole, a typical type of Chi-pao at that time.

襟：衣襟从领口直通衣摆。
Front Piece: The front piece is the part from the collar to the hem.

"团龙纹"图案整齐划一，十分精美。
The Round Dragon Pattern is neat, complete, and extremely beautiful.

袖口绣有蝶恋花图案。
The cuffs have the Butterfly Loving the Flower patterns.

• 清代的典型旗袍样式

旗袍是清代满族女子日常穿的袍服。这件旗袍用料考究，做工精细，色泽艳丽，衣身宽大，线条平直，是典型的清代贵族女子服饰。

A Typical Chi-pao Style in the Qing Dynasty (1616-1911)

Chi-pao is a daily gown for Manchu women in the Qing Dynasty (1616-1911). This Chi-pao is made of the first-class materials and showed refined tailoring. With bright colors, a loose design and straight lines, it is a typical dress for noble women in the Qing Dynasty.

- 油画《江南女子》（作者：姜迎久）

清代中期以后，满族和汉族女性的服装互相影响并仿效，发生了很多细微的变化。比如，旗袍袍身从宽大变得紧身，袖子由窄而紧的马蹄袖变得宽大，领口处还被加了一寸多高的立领。

The Canvas *a Girl from South of the Yangtze River* (by Jiang Yingjiu)

After the middle period of the Qing Dynasty, the Han and Manchu women's dresses influenced and imitated each other and underwent many minute changes. For example, Chi-pao changed from tight to loose and its sleeves were much wider than the previous narrow ones in horse-hoof shape. A standing collar was added.

"大拉翅"与"花盆底"

清代的满族女性非常注重头饰和发型，一般会梳高髻，流行戴假发髻，装饰在脑后。到了清代中后期，满族女性的发型更加高耸，逐渐演变成为一种被称作"大拉翅"的发髻。清代的汉族女性并不注重发型和头饰，发型简约，发髻低矮、伏贴，给人含蓄、谦恭的感觉。

清代满族女性穿的鞋子通常很高，鞋底中部设有木制高底，形状像一个花盆，俗称"花盆底"。这种"花盆底"踩在地上很像马蹄印，所以又被称为"马蹄底"。也有一些满族女性喜欢穿轻便的绣花鞋。

- 清代满族女子的"大拉翅"发饰

清末满族贵妇头戴"大拉翅"，中间镶嵌着一朵牡丹花，右前额缀满了小花。

The Big Wing Hair Decoration of Manchu Women in the Qing Dynasty (1616-1911)

By the end of the Qing Dynasty, noble women often wore Big Wing headwear. They put a peony in the middle and some small flowers on the right forehead.

Big Wing and Vase Bottom

Manchu women in the Qing Dynasty took their headwear and hairdo seriously. Normally, they wore their hair in high buns and often put artificial buns on the back of their heads. In the middle and late periods of the Qing Dynasty, Manchu women made their buns even higher and created the Big Wing bun. However, Han women in the Qing Dynasty took their hairdo and headwear lightly. They favored simple and low buns, which gave a reserved and humble feeling.

Manchu women in the Qing Dynasty wore loose shoes with a high wooden sole in the middle. Shaped as a vase, it was called the Vase Bottom. Because it would make a horse-hoof print on the ground, it was also called the Horse Hoof Bottom. Of course, light embroidered shoes were also liked by some women.

- 清代汉族女性的日常装扮

Daily Attire of Han Women in the Qing Dynasty (1616-1911)

端正花：位于头板正中的彩色大型绢花。端正花的周围，即头板的其他位置通常被加饰以小鲜花及其他首饰。

Straight Flower: It is a large colorful silk flower in the center of the Head Board. It is surrounded with small flowers and other ornaments.

头座：底部呈扣碗状，以铁丝制成。做头饰时，将头座扣于发顶圆髻上固定，将多余的发丝缠绕其上，使头座不被看出。

Head Pedestal: Its bottom is like a bowl placed upside down. It is made of iron wire. When making the headwear, the Head Pedestal is fastened onto the hair bun and covered with hairs.

头板：为"不"字形，多以黑色的绸缎、绒布、纱制成。

Head Board: The Chinese character "不"-shaped board was mainly made of black silk, cotton flannel and gauze.

燕尾式发髻：后颈的发髻呈燕尾式。

Swallow Tail Bun: The bun on the back side of head has a swallow-tail shape.

顶髻：在发顶梳成圆髻，以便头座扣于其上。

Top Bun: There is a round top bun for fastening the Head Pedestal.

- 清代满族贵妇的日常装扮
Daily Attire of Manchu Noble Women in the Qing Dynasty (1616-1911)

- 花盆底鞋
Vase Bottom Shoes

> 旗袍的演变

旗袍的变化在近百年来体现得非常明显，变得风情万种。这不仅仅是一种服装形式的变化，更体现着一种文化的发展和成熟。随着人们生活方式和审美情趣的变化，旗袍演绎出多姿多彩的样式，让人目不暇接。

> Evolution of Chi-pao

Chi-pao has undergone huge changes in the last century. It has become exceedingly fascinating and charming. These changes are not only in form, but more importantly, they represent cultural development and maturity. With the change in people's lifestyles and aesthetic interests, Chi-pao has evolved into diversified styles.

At the end of the 19th century and the beginning of the 20th century, Western clothing was spread to China together with advanced production technologies. In many coastal cities, especially the international metropolises such as Shanghai, the clothing has had obvious changes. In the first decade of the 20th century, the lifestyle

- 油画《素弦声歇》（作者：刘文进）
  The Canvas *a Beauty in Performance* (by Liu Wenjin)

19世纪末20世纪初,西式服装和先进的生产技术开始传入中国,在很多沿海大城市,尤其是像上海这样的国际化大都市,服饰发生了明显的变革。在20世纪初期的十多年里,中国人的生活方式和衣着观念进入了新旧更替的时期,穿旗袍的人大大减少,旗袍在民间不再流行,大多为满族贵族家庭所独有。

不过,此时的旗袍仍悄然发生着变化。虽然整体上仍然承袭着清代晚期的宽身造型,线条也很平直,但受当时崇尚自然的社会风气影响,旗袍的样式逐渐被简化。袖子收紧并缩短,露出手腕,袍身长度缩短至脚踝处;在装饰上也被大刀阔斧地删繁就简,镶滚、刺绣等装饰减少,色调也变得淡雅,整体风格趋于简约、素雅。

and dressing concept of the Chinese people were changing drastically. Fewer and fewer people chose to wear the Chi-pao. Only the noble Manchu families kept on wearing it.

However, the Chi-pao at that time was still changing quietly, although its overall contour was still the loose design with straight lines popular in the late Qing Dynasty. Influenced by the social tendency that esteemed natural elements, the Chi-pao was gradually simplified. Its sleeves were tightened up and shortened to expose the wearer's wrists. Its hem was lifted to ankles. Its adornments such as lacing and embroidery were greatly reduced and it had lighter colors. The overall style tended to be simple and elegant.

- **满族女性的冬季装束(19世纪末20世纪初)**
  画面上的女子梳着经典的两把头发式,前额没有刘海,穿着一件高领棉旗袍。
  Winter Attire of Manchu Women (End of the 19th Century and Beginning of the 20th Century)
  The woman in the picture wears a typical two-half hairdo without bangs on her forehead. She is in a high-necked cotton Chi-pao.

- **满族宫廷贵妇服饰（19世纪末20世纪初）**

照片上的女性服饰仍为传统的旗装，但高耸的"大拉翅"发型已经不见了，旗袍的袖口变得宽大。

Attire of Manchu Noble Women (End of the 19th Century and Beginning of the 20th Century)

The attire in the picture is still the traditional Chi-clothing. However, the high Big Wing hairdo has disappeared. The cuffs of the Chi-pao are loose.

- **清朝最后一位皇后婉容像**

Wanrong, the Last Empress of the Qing Dynasty

上衣下裤是当时比较流行的服装样式，袖子又窄又紧，短至肘部，袖口处露出方格衬衫的袖口。一条长纱巾很有特色，经头顶缠绕下来，垂至膝部。这在当时是很时髦的装扮。

Coat and trousers were the popular dressing form at that time. The sleeves were narrow and tight and covered no further than the elbow. The cuffs of the plaid shirt protruded out. A signature long gauze scarf wound its way down from the top of the head to the knee. It was a fashionable dressing style.

这位女子也披了一条长纱巾，在浅色的短袄外加罩了一件长及脚面的坎肩，领边滚以两道醒目的花边。短袄的袖口较紧，镶有精细的花边，一段白色的衬衫袖子露了出来，精致而素雅。

This woman had a long gauze scarf. She wore a long waistcoat over a light-colored short coat. The collar had two eye-catching laces. The short coat had tight cuffs with exquisite lace. A section of the white shirt sleeves was exposed, showing an elegant and graceful dressing style.

这位女子的衣着更为简洁，上袄下裙，衣身略有些宽松，线条平直。头戴饰花发带，胸前佩有珠串，体现着当时很时髦的装扮。

Her dressing was even simpler. She wore a coat and a skirt, which were loose with straight lines. She had a flowery hair band on her head and a rosary on her chest. This was a fashionable dressing style at that time.

此女子侧身坐在栏杆上，上衣下裤，元宝领几乎遮盖了半边脸颊，高领下佩戴着夸张的花饰。

This woman sat sideways on the handrail. She wore a coat and trousers. Her shoe-shaped collar nearly covered half of her face. Under the high collar were exaggerating flower decorations.

"元宝领"的出现和广泛应用是这一时期旗袍的重要特色之一。元宝领为立领，高达四五厘米甚至六七厘米，盖住腮部，几乎齐耳，从正面看形如元宝。

The advent and wide use of the shoe-shaped collar was one of the important features during that period. The shoe-shaped collar stood up as high as four or five centimeters, or even six or seven centimeters. It covered the cheeks and nearly reached the ears. Viewed from front, it had a shoe shape.

- 20世纪初的新式高领女装

照片中这四位女子一字排开，或坐或立，有的着裙装，有的着裤装，各有特色。上衣的领子高高竖起，是当时流行的着装形式。

The New Style of High-Necked Women's Dresses in the Early 20th Century

The four women in the picture are in various attires, some in skirt and some in trousers. Their coats have high collars, a fashionable design at that time.

1920年前后，中国掀起了"五四"新文化运动，人们开始接受一些新的思想观念，也促进了女性的身心解放。在当时的北京、上海、广州、南京等地，女性追求时髦、新潮已蔚然成风。因此旗袍呈现出多样化的发展，成为时尚服饰的主角。

这一时期的旗袍在保留自身特

Around 1920, the May Fourth New Culture Movement took place in China. The Chinese people began accepting some new ideas and concepts, which promoted the emancipation of women in both mind and body. At that time, women in places such as Beijing, Shanghai, Guangzhou and Nanjing closely followed fashion and new trends. Therefore, the Chi-pao underwent diversified

"五四"新文化运动

"五四"新文化运动简称"五四运动"，是1919年5月4日在北京爆发的中国人民彻底的反对帝国主义、封建主义的爱国运动，冲击了统治中国两千多年的封建专制制度和传统观念，对中国的思想文化，政治发展方向，社会、经济潮流都产生了重大的影响。

May Fourth New Culture Movement

Also known as the May Fourth Movement, the May Fourth New Culture Movement broke out on May 4th, 1919 in Beijing. A patriotic movement of the Chinese people against imperialism and feudalism, it dealt a heavy blow to the feudal dictatorship and traditional ideas that had dominated China for over two thousand years and had a significant impact on China's ideology, culture, politics, and social and economic trends.

• **女学生的"文明新装"**

受日本服饰风格的影响，当时在中国的城市时兴过一阵改良的"文明新装"，上为短袄，下配黑色长裙，多由接受过新式教育的女教师和女学生率先穿着，因朴素而大方、清纯而淡雅，很快被城市女性引为时尚，纷纷效仿。

New Civilized Dress for Girl Students

Influenced by Japanese clothing style, Chinese girls in cities at that time once favored the modified New Civilized Dress, which consisted of a short coat and a black long skirt. The Dress was first accepted by female teacher and girl students who received modern education. Before long, it was popular among women living in cities due to its simple and tasteful style and light and elegant design.

色的同时，还引进了一些西式时装的元素，如在袖口设计上采用西式风格，在领、袖、襟、下摆等衣缘处加入欧洲的时髦元素，常见的有锯齿形、圆齿形、波浪形等。当时，穿进口丝袜开始在上海的职业女性中流行。丝袜是对传统布袜的改革，使女性更加妩媚动人。高跟鞋与丝袜搭配成为身着旗袍的女子的时髦装扮。

这时的旗袍的外型开始呈现出收腰之势，腰线较低，胸部、腰身、臀部的曲线略突出；虽仍为倒

development and became the leading attire among fashionable dresses.

During that period, while keeping some of its characteristics, Chi-pao adopted many elements of Western fashionable dresses. For example, it had Western-style cuffs. On the fringes of its collar, cuffs, front piece and hem were vogue European patterns such as sawtooth, knuckle tooth and waves, which were popular among the career women in Shanghai. The silk stockings were a reform to the traditional cloth socks. The stockings added charms to

- **身穿马甲旗袍的女子（20世纪20年代）**

  20世纪中叶，马甲旗袍风靡一时。马甲旗袍衣身宽松，线条平直。马甲长及足背，与短袄合为一体，省去了上身的重叠部分，袖子为倒大袖式，在领、襟、摆等部位采用镶滚装饰，成为新式旗袍的雏形。画面上的女子轻倚栅栏，身着马甲旗袍。宽松的袍身掩盖不住青春气息，淡雅的色调与身后的山水相呼应，清新而雅致。

  A Woman in a Vest Chi-pao (1920s)

  In the middle of the 20th century, vest Chi-pao was popular. Its upper part was loose with straight lines. The vest reached the wearer's feet and it integrated with the short coat. The upper overlapping part was omitted. The reversed large sleeves were adopted. Edging decorations were added to the collar, front piece and hem, forming the embryonic shape of the new-style Chi-pao. The girl in the picture sits on the handrail. Her vest Chi-pao is loose but can hardly conceal her youthful vitality. The light and elegant color of the Chi-pao fits in well with the landscape around her. It is indeed a fresh and graceful scene.

大袖，但袖口变小，装饰也趋于简洁；下摆逐渐变短，摆线提高至膝盖处，露出了美丽的小腿。此外，旗袍的布料种类更加丰富，有绒、

- **穿旗袍的母亲和孩子们（20世纪20年代）**
照片中的这位年轻母亲短发齐耳，戴眼镜，穿着当时流行的小高领旗袍，侧领系扣，刚好遮住膝盖，露出浅色的丝袜，皮鞋也同样是浅色的，清新而雅致。男孩子们穿着具有西式风格的翻领衬衣。女孩儿戴着一顶钟形帽，帽子的右侧还配有花穗，显得活泼而可爱。

A Mother in Chi-pao and Her Children (1920s)
The young mother in the picture has a short hairdo. She wears glasses and a high-necked and side-buttoned Chi-pao popular at that time. The hem of the Chi-pao just covers her knees and exposes her light-colored silk stockings. Her shoes are also light-colored, giving her a fresh and elegant look. The boys wear Western-style shirt with turndown collars. The girl wears a bell-shaped hat decorated with some flowery tassels on the right. She looks cute and adorable in her clothes.

their wearer's. The high-heeled shoes and silk stockings were the articles indispensable to the women wearing Chi-pao.

At that time, Chi-pao began to have a tightened and lowered waist line and more highlighted curves on the chest, waist, and hips. Although it kept the reversed large sleeves, the cuffs dwindled and there were simpler adornments. The hem became shorter and was lifted to above the knees, exposing women's beautiful shanks. In addition, Chi-pao was made of more diversified materials, including new ones such as flannel, woollen material, gauze and lace.

此女子上穿明黄色梅花纹短袄，下着白色几何纹褶裙，倒大袖的边缘被设计成波浪形，裙摆也是如此，非常新颖、别致。

This girl wears a bright yellow coat with plum blossom patterns and a white pleated skirt with white geometric patterns. Both the reversed large sleeves and the skirt hem have wave-like fringes, a novel and elegant design at that time.

此女子身穿一件浅色旗袍，采用带有菊花纹的进口面料，领口、袖口、衣襟处被镶滚了两道花边，简洁而大方。

This girl wears a light-colored Chi-pao made of the imported materials with chrysanthemum patterns. Its collar, cuff and front piece have two laces, guaranteeing a simple and tasteful impression.

- **湖畔女子（20世纪20年代）**
  这一时期的旗袍带有西式时髦元素的边缘设计显得富有变化，多姿多彩。画面上的两位年轻女子所穿的服装体现着当时的流行风尚。

  Beautiful Girls by a Lake (1920s)
  The Chi-pao during that period had the fashionable Western-style fringe design, which made the dress for women full of changes. The dresses of the two young ladies in the picture reflect the fashion at that time.

呢、纱、蕾丝等各种新式面料。

20世纪三四十年代是旗袍发展的全盛阶段，最终形成了现代女性所穿的新式旗袍，奠定了旗袍在中国女性服装史上的经典地位。当时，上海的贵妇、名媛和各类时髦女子在各种场合尽情展示着旗袍的风姿，一大批当红电影明星都对旗袍情有独钟，在银幕上下演绎着旗

The 1930s and 1940s saw the full-swing development of the Chi-pao, which finally gave birth to the new-style Chi-pao worn by modern women and laid the foundation for Chi-pao to become a classical dress for women in Chinese clothing history. At that time, ladies in high society, and women of fashion in Shanghai wore Chi-pao on various occasions and fully demonstrated

袍的万千风情，如胡蝶、阮玲玉、周璇等。

　　这一时期的旗袍款式有两大特点，一是中西合璧，二是变化多端。旗袍被不断进行着工艺改进，进入了立体造型时代，旗袍的衣身有了前后片之分。尤其到了20世纪40年代后期，中国的时装业已非常

its charms and merits. A large group of movie stars had partiality for Chi-pao and they displayed its diversified charms both on and off the screen. Among them were Hu Die, Ruan Lingyu and Zhou Xuan.

　　During that period, the design of Chi-pao had two major characteristics: the combination of Chinese and Western Elements and endless changes. Through nonstop improvement, Chi-pao entered the era of 3D modeling. Its main part evolved into two pieces: the front piece and the back piece. By the late 1940s, the Chinese fashion industry had been highly developed. New dresses and new designs emerged one after another. The clothing styles were even more diversified. Women's dress became more delicate, charming and prettier. Influenced by such a trend, Chi-pao began its evolution towards a fashionable dress. Some new

- 油画《阮玲玉》（作者：刘文进）
阮玲玉是20世纪30年代中国著名的影星，主演过近20部影片。
The Canvas *Ruan Lingyu* (by Liu Wenjin)
Ruan Lingyu was a famous Chinese movie star in the 1930s. She starred nearly 20 movies.

发达，新服装、新样式层出不穷，风格更加多样，女性服装则更加娇艳、妩媚动人。在这种风潮的影响下，旗袍也开始走向时装化，一些新样式涌现出来，比如旗袍裙。时髦的女性还常在修长而收腰的旗袍外穿裘皮大衣、西式外套、背心等，再配上烫发、丝巾、丝袜、项链、耳环、手表、皮包、高跟皮鞋等，体现着当时最时尚的装扮。

designs, like the Chi-pao skirt, emerged. Women of fashion often wore a fur coat, a Western-style coat, or a vest over the long Chi-pao with a tightened waist. They also had their hair permed and adopted adornments such as silk scarves, silk stockings, necklaces, earrings, watches, purses, high-heeled shoes, and so on. These were the fashionable adornments for a Chi-pao wearer at that time.

In the middle of the 20th century,

- **上海的集体婚礼（20世纪30年代）**
  在这场集体婚礼中，男士统一着蓝袍、黑褂，女士穿粉色短袖长旗袍，头罩白纱，戴白手套，手持鲜花。
  A Group Wedding in Shanghai (1930s)
  In the wedding, the gentlemen wore blue robe and black gown while the ladies wore pink full-length Chi-pao with short sleeves, each with a white gauze covering her head. They also wore white gloves and held a bunch of flowers.

- **胡蝶着旗袍的照片**

  胡蝶是20世纪三四十年代中国著名的影星，气质高贵而典雅，一度被观众评为"电影皇后"。

  Hu Die in Chi-pao

  Hu Die was a famous Chinese movie star in the 1930s and 1940s. Known for her noble temperament and graceful beauty, she was once dubbed Movie Queen by the audience.

- **民国女性着旗袍行走在街上的照片**

  此时的旗袍引入了许多欧美流行元素，如加入了拉锁、按扣、垫肩等小配件。材质是纱、绸、缎、棉等；各种新颖的纺织品也从欧美被引进，这些面料质地柔软，手感好而且富有弹性，成为都市女性制作旗袍的新选择。还有一些透明和镂空的化纤或丝织品，常被做成衬裙穿在旗袍里。

  Women in Chi-pao Walking in the Street During the Republic of China (1912–1949)

  During that period, Chi-pao was seen with many popular elements imported from Europe and America, such as zippers, press buttons and shoulder padding. Collar was made of all conceivable materials, including gauze, silk, satin, cotton, and various new fabrics imported from Europe and America. Soft, elastic, and touching well, these materials were new options for women in the cities to make their favorite Chi-pao. Some transparent and hollowed-out chemical fibers or silk fabrics were often used to make underskirts.

- **女性夏季服饰（20世纪40年代）**

  除了长旗袍，短旗袍也流行过一段时间，低开衩，以仅露小腿为尚。女子身着这种旗袍，坐、立、行走的姿态自然、含蓄、优雅。画面中的女子倚树而立，身穿一件蓝色短旗袍，显得清新而自然；裙摆只到膝盖处；袖口仅盖住肩部，露出两只胳膊，洋溢着青春气息。

  A Woman in Her Summer Attire (1940s)

  Besides full-length Chi-pao, short Chi-pao had also been in vogue for a while. Its low slits exposed the wearer's shanks only. Women wearing such Chi-pao showed reserved and graceful bearing during sitting, standing and walking. The woman in the picture stands by a tree. Wearing a blue short Chi-pao, she has a fresh and natural manner. The hem of the Chi-pao, reaches only her knees. The cuffs cover her shoulders only and her arms are totally exposed. The dress fully displays her youthful vitality.

- **身穿无袖长旗袍的女人（左）**

  抗日战争胜利后，华美、考究的旗袍再度盛行，新式旗袍的改良程度加大，更加明显的省道突出了女性的曲线美。旗袍的面料多样，各式各样的装饰被广泛运用，与披风、披肩、西式的帽子等搭配，呈现出明显的时装化趋势。

  The Woman in a Sleeveless Full-length Chi-pao (Left)

  After China's victory in the Anti-Japanese War, the well-tailored Chi-pao again took the center stage. New-style Chi-pao showed greater modification. Its more obvious dart lines accentuate the line of beauty of women. Chi-pao at that time was made of various materials. The decorations were used together with mantle, shawl and Western-style hat, giving rise to an obvious trend towards fashionable dressing.

025 Origin of Chi-pao
旗袍的由来

- 身穿长旗袍的女人（20世纪30年代）

长旗袍一度成为20世纪30年代时髦女性的标准服装。画面中的女子身材窈窕，旗袍款式新颖，衣摆长可及地，身上的花纹体现了裁剪的精妙，脚上穿着当时非常流行的鱼嘴鞋，从衣摆下微微露出。

A Woman in a Full-length Chi-pao (1930s)

Full-length Chi-pao was once a standard dress for women of fashion in the 1930s. The woman in the picture has a graceful figure. Her Chi-pao is a new design, with a hem almost touching the ground. The flowery pattern fully demonstrates the refined tailoring. She wears a pair of fashionable fishmouth shoes, which vaguely show up beneath the hem.

月份牌中的旗袍

月份牌是20世纪初盛行的一种广告宣传画，类似于中国传统年画的形式，画面上标有商品、商号和商标，并配以中西对照的年历或西式月历。旗袍时装美女是月份牌最常见的形式，身穿旗袍的清纯的女学生和电影红星经常在月份牌中出现，因而月份牌又俗称"美女月份牌"。

The Chi-pao in Monthly Poster

The monthly poster was an advertising picture popular in the early 20th century. Similar to China's traditional New Year pictures, it carried information about the commodity, its distributor and trademark, accompanied by solar and lunar calendars or Western-style monthly calendar. Beauties in Chi-pao were the commonest images on monthly poster. The beauties were mainly innocent girl students and movie stars. Therefore, the monthly poster was also called the Beauty Monthly Poster.

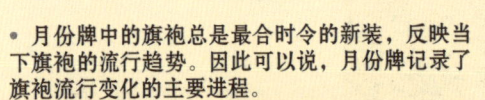

- 月份牌中的旗袍总是最合时令的新装，反映当下旗袍的流行趋势。因此可以说，月份牌记录了旗袍流行变化的主要进程。

  The Chi-pao shown in the monthly poster was always the latest design and represented the most popular trend at that time. Therefore, the monthly poster recorded the mainstream evolution of Chi-pao during that period.

20世纪中期，旗袍的形式逐渐简化，形成一种固定的模式——立领、长身、收腰、低开衩；很少使用刺绣、镶滚等装饰工艺，以织物图案代替了繁复的装饰；用色单纯，色调素雅而和谐；面料多为中、低档，高档面料的旗袍则多用于进行外事活动。50年代末，旗袍逐渐被中山装、人民装代替。到了60年代，旗袍基本在中国大陆销声匿迹了。

Chi-pao was gradually simplified and frozen into a fixed style, which featured a standing collar, a long design with contracted waistline, low slits, and little decorations such as embroidery and lacing. The fabric patterns replaced the complicated adornments. Its colors were pure, simple, elegant, and harmonious. Medium and low-grade materials were mostly used. The Chi-pao made of high-grade materials was mainly for diplomatic events. By the end of the 1950s, Chi-pao had been substituted by Chinese tunic suits or people's suits. By the 1960s, it totally disappeared in the mainland of China.

However, during the two to three decades when Chi-pao was given a cold

- 端庄而典雅的旗袍女人（20世纪50年代）

这一时期的旗袍风格讲求简洁、大方、立领、方襟、带盘扣、衣袖合体，低开衩，胸口绣有花卉图案，使穿着者具有一种不妖、不媚、不纤巧的姿态。画面中的女子身穿一件缎面旗袍，胸前佩有贴花装饰，搭配一串珍珠项链，仪态淑雅而端庄。

A Graceful Woman in Chi-pao (1950s)

The Chi-pao during this period featured a simple and tasteful style. It had a standing collar, a square front piece, some frogs and low slits. The main part of the Chi-pao and the sleeves formed an integral whole. There were embroidered flower patterns on the chest, which gave the wearer a look not coquettish, flattering or manipulating. The woman in the picture wears a satin Chi-pao. She has appliqué decorations on her chest, accompanied by a pearl necklace. In such a dress, she looks graceful and dignified.

● 1948年上海街头广告牌中的旗袍女人
A Chi-pao Beauty on a Street Poster in Shanghai, 1948

然而，在旗袍在中国被冷落的二三十年里，海外成为它发展的舞台。旗袍以离地约20厘米为标准，设计趋向曲线化，裙摆略小，更加注重腰身的曲线美，深受广大女性的喜爱。在此期间，旗袍曾多次被当成作品参加国际服装展，并屡屡获奖，受到了很多世界著名服装设计师的赞誉，同时以其独特的魅力影响着他们的设计。由于旗袍在国外的知名度日渐提高，很多国外的名人和影星来中国定制旗袍。

20世纪80年代初，旗袍在中国大陆再度崛起，成为中国女性在正

shoulder in China, it was developed outside China. The hem of Chi-pao should be 20 centimeters above the ground and its design features line of beauty. With a small hem, it highlights women's beauty and is loved by women. Back then, Chi-pao entered many international fashion shows and always won a prize. It was highly praised by many world famous fashion designers and it influenced their design with its unique charm. As Chi-pao enjoys greater and greater fame in the world, many celebrities and movie stars from other countries come to China to have their Chi-pao tailor-made.

式场合穿着的礼服。1989年，苏州刺绣厂设计并绣制的一件旗袍在首届北京国际博览会上获得了金奖。但在大多数人眼中，旗袍显得不太合时宜，因而只有少数人穿着。

从20世纪90年代中期开始，女性以身材高挑而纤长、平肩、窄臀为理想形象。旗袍这种最能衬托东方女性身材和气质的服装又重新为人们所重视，受到很多女性的欢迎，尤其成了社交场合的礼服。

与此同时，世界服饰舞台上更是刮起了中国风，中国的龙凤吉祥

In the early 1980s, Chi-pao rose again in the mainland of China and became the dress for women on formal occasions. In 1989, a Chi-pao designed and made by Suzhou Embroidery Factory won the gold award at the first Beijing International Exposition. However, most people regard Chi-pao as out of date and only a few are willing to wear it.

From the mid 1990s, the ideal female stature has been a tall and thin one with flat shoulders and narrow hips. As Chi-pao can best showcase the stature and temperament of women in the East, it has again been treasured by people and loved by women. It is accepted as the formal attire for social gatherings and diplomatic events in China.

Meanwhile, a China current is going strong on the world fashion stage. The auspicious Chinese patterns of loong and phoenix and Chinese character images are all viewed as fresh design elements. The arc hem, frog, and standing collar of Chi-pao are adopted by Western fashion designers in their design and this

- 油画《窗前无语》（作者：刘文进）
 The Canvas *a Quiet Lady by the Window* (by Liu Wenjin)

图案及文字形象都被视为新鲜的设计元素。旗袍的圆弧形下摆、盘扣、立领等都被西方设计师应用在时装设计上，出现了时装化的旗袍或旗袍风格的时装。总之，旗袍作为中国文化的代表性符号之一，优美的曲线、精美的刺绣和盘扣、高贵的立领等，开始被西方人吸纳为设计的元素。

经过300多年的发展，旗袍已今非昔比，可谓被进行了颠覆性的改良，无论在裁剪上还是用料上，都发生了前所未有的变化。现在的旗袍已不再拘泥于传统的样式，

gives rise to the Chi-pao with fashion elements or the fashionable dresses with Chi-pao features. In a word, as one of the representative symbols of Chinese culture, Chi-pao has many unique features that are accepted by Westerners as design elements, including its graceful; lines, exquisite embroidery and frogs, noble standing collar, etc.

Three centuries after its advent, Chi-pao has now reached a great height after some subversive modifications. In both tailoring and material application, it has gone through unprecedented changes. Current Chi-pao is no longer confined

外国服装设计师对旗袍元素的运用

近年来，很多外国著名时装设计大师也经常在作品中融入旗袍元素，国际时装舞台上劲吹中国风。一些国际顶级设计师和国际大品牌，如阿玛尼、夏奈尔、皮尔·卡丹等都越来越频繁地将中国传统文化元素融入服装设计中，法国服装大师皮尔·卡丹也曾承认自己从中国旗袍中获取了大量灵感。在2011年的路易威登巴黎时装秀上，旗袍元素更是被运用到了极致。

Application of Chi-pao Elements by Foreign Designers

In recent years, many master fashion designers in other countries have often adopted Chi-pao elements in their design. A China current has been going strong on international fashion stage. World top designers and top international brands such as Armani, Chanel, and Pierre Cardin have been more and more frequently merged traditional Chinese cultural elements into fashion design. French master fashion designer Pierre Cardin once admitted that he had been greatly inspired by Chinese Chi-pao. During the 2011 Louis Vuitton Fashion Show in Paris, Chi-pao elements were put to the finest use.

露肩、露脐、露腿，甚至露胸的旗袍被众多时尚女性喜爱并追求。此外，很多旗袍设计中都被加入了新元素，使旗袍步入高档时装的行列。2008年北京奥运会颁奖典礼上的礼仪小姐均穿旗袍，将东方女人的美丽动人表现得淋漓尽致，令很多外国友人大加赞叹。

现代旗袍与传统旗袍早已有了非常大的区别，比如衣长、袖长普

to traditional style, but dares expose the wearer's shoulder, navel, leg, and even breast. Such design is wooed by countless ladies of fashion. In addition, many Chi-pao designs contain various new clothing elements, which put them among first-class fashions. In the 2008 Beijing Olympic Games, all the ritual girls on the award-giving ceremonies wore Chi-pao. The dress fully demonstrated the beauty and charm of Eastern women. Many foreigners highly praised the beauty of the Chi-pao.

Greatly different from traditional Chi-pao, modern Chi-pao has shorter main piece and sleeves. Its sleeves are mainly in Western style. Traditional Chi-pao had a front piece in many shapes, including *pipa* (a four-stringed Chinese lute) shape, *ruyi* (an S-shaped ornamental object symbolizing good luck) shape, and oblique shape. However, it was mainly in an asymmetric design with buttons arranged on the right, a traditional habit with Chinese clothing. Since the ancient

- **现代时装旗袍**
这件旗袍的面料是红色织锦缎，大花盘扣打破了传统盘扣的样式，起到了画龙点睛的作用。
A Modern Chi-pao
This Chi-pao is made of red tapestry satin. Its large flowery frog is different from the traditional style and better highlights the beauty of the dress.

遍缩短，袖子也主要以西式袖为主。再如，传统旗袍的襟形很多，如琵琶襟、如意襟、斜襟等，但一般都是非对称的右开襟，体现了中国传统的习惯。因为中国自古以来就有崇右的思想，所以中国古代服饰大多为右开襟。而西方人在设计斜开襟的服装时，几乎全部是左开襟，这一习惯也影响到了现代旗袍的设计，使改良旗袍的款式更加丰富。

最近几年，中国很多服装设计师也在努力改进旗袍与现代生活不相适应的弱点，设计了很多新式旗袍。这些旗袍款式继承了传统旗袍的优点和特征，又与现代生活结

times, the Chinese people have highly esteemed right. That was why most of the ancient Chinese arranged the buttons of their clothes on the right. When Westerners design the clothes with an oblique front piece, they put buttons on the left, too. This habit has also affected the design of modern Chi-pao and enriched the styles of the modified Chi-pao.

In recent years, many Chinese fashion designers have been trying hard to make Chi-pao more suitable for the modern lifestyle. They have designed many new styles of Chi-pao, which not only inherit the merits and features of the traditional Chi-pao, but better fit in with modern life. Such new Chi-pao

● 现代旗袍秀（图片提供：CFP）
Modern Chi-pao Show

合，更加现代化、实用化，更符合现代人的审美趣味。可以说，改良后的旗袍已经开始走入人们的日常生活，并不只是宴会、婚礼等特殊场合的专用服饰了。

当然，无论是经典的款式，还是前卫的新样式，旗袍还是旗袍，都能衬托出女人的韵味，呈现女人妩媚而高雅的美感。

is more practical and better meets the aesthetic taste of modern people. In this sense, the modified Chi-pao has begun entering people's daily life. It is no longer a special attire for formal occasions such as banquets and weddings.

Of course, in either classical style or fashionable design, Chi-pao is still Chi-pao and it is always ready to showcase the charm and elegance of women.

- **加入现代元素的旗袍**

  这款旗袍汲取了西式礼服的元素，前身是中国风格的刺绣荷花图案，后身以真丝营造出朦胧的感觉，使肌肤若隐若现，下摆也采用了中国盘扣的装饰，在整体上给人以华丽、庄重、大方之感的同时，又有浓厚的中国气息。

  A Chi-pao with Modern Elements

  This Chi-pao contains some elements of Western formal attire. Its front piece bears the embroidered lotus flower patterns in Chinese style. Its real silk back creates a hazy atmosphere and makes the skin partly hidden and partly visible. Its lower part also has Chinese frog decoration. As a whole, it looks gorgeous, solemn and tasteful, together with a strong sense of Chinese characteristics.

旗袍的发展和演变
The Development and Evolution of Chi-pao

清初的满族旗袍 Manchu Chi-pao in the early period of the Qing Dynasty	圆领，袖子窄而紧，领口与大襟处有细滚边；衣身略窄，衣摆呈喇叭形，没有开衩。 Round collar, narrow and tight sleeves, fine edging at collarband and front piece, narrow body part, horn-shaped hem without slit.
清中期的满族旗袍 Manchu Chi-pao in the middle of the Qing Dynasty	更加宽而肥。小立领，大襟；衣袖宽而肥；直身，圆下摆。 Wider and fatter, small standing collar, large front piece, loose sleeves, straight body part, round hem.
清后期的满族旗袍 Manchu Chi-pao in the late period of the Qing Dynasty	衣身已经变得十分宽而肥，仍保持着直身的外形；圆领，大襟，圆下摆，两侧开衩。 Very wide and fat body part, still with a straight contour, round collar, large front piece, round hem, slits on both sides.
19世纪后期的汉族妇女服饰 Attire of Han women in the late 19th century	上衣为大袄，元宝领，直身，窄袖，衣袖短至手腕，衣身开衩。下身为百褶裙，裙上带有如意云头纹。 A large coat with shoe-shaped collar, straight body part and narrow sleeves, cuffs shortened to wrists, slits in body part; a pleated skirt with auspicious cloud patterns.
20世纪20年代汉族女子的服饰 Attire of Han women in the 1920s	上为短袄，立领，大襟，窄身，倒大袖，圆下摆，衣领、衣襟、侧襟均有盘扣。下身为黑裙，长及脚面。 A short coat with standing collar, large front piece, narrow body part, reversed large sleeves, round hem, and frogs on collar, front piece and side piece; a black skirt touching the feet.
20世纪20年代的马甲旗袍 Vest Chi-pao in the 1920s	当时流行在短袄外面加罩一件长马甲，取代长裙，这种款式被称为"马甲旗袍"。 A long vest over the short coat replacing the long skirt, a popular style at that time, known as vest Chi-pao.

20世纪20年代的倒大袖旗袍 The Chi-pao with reversed large sleeves in the 1920s	立领，方圆襟，衣袖为典型的倒大袖，衣长至小腿，无开衩，略显腰身。 Standing collar, square and round front piece, typical reversed large sleeves, hem reaching shanks, without slit, body part revealing the figure.
20世纪20年代的长袖旗袍 Long-sleeve Chi-pao in the 1920s	立领，斜襟，衣袖窄而直；盘扣分布在领口、斜襟、侧门襟处；收腰明显，两侧开衩。 Standing collar, oblique front piece, narrow and straight sleeves, frogs on collar, front piece and side piece, obviously contracted waist, slits on both sides.
20世纪30年代的短袖旗袍 Short-sleeve Chi-pao in the 1930s	立领，斜襟，短袖；窄身，衣摆至小腿，两侧开衩；盘扣分布在领、襟、侧门襟处。 Standing collar, oblique front piece, short sleeves, narrow body part, hem reaching shanks, slits on both sides, frogs on collar, front piece and side piece.
20世纪30年代的扫地旗袍 Ground-sweeping Chi-pao in the 1930s	立领，斜襟，无袖；收腰，直下摆，长及地面，两侧开衩较低。能衬托出女性纤长的身材。 Standing collar, oblique front piece, sleeveless, contracted waist, straight hem reaching the ground, low slits on both sides, demonstrating women's thin and slender body.
20世纪40年代的短旗袍 Short Chi-pao in the 1940s	此时的旗袍制作出现了省道结构，胸省、腰省、臀省使旗袍更加贴身，线条简洁、生动。 The Chi-pao of this period was provided with dart structures, such as chest dart, waist dart, hip dart, which made Chi-pao much fitter with a simple and lively line.
20世纪40年代的长袖旗袍 Long-sleeve Chi pao in the 1940s	高领，斜襟，在衣襟处使用了按扣，衣袖长而窄，袖笼收省，肩部安有垫肩；收腰明显，两侧开衩。 High collar, oblique front piece, buttons on front piece, long and narrow sleeves, tightened up cuffs, with shoulder pads, obviously contracted waist, slits on both sides.

20世纪40年代的无袖旗袍 Sleeveless Chi-pao in the 1940s	立领，斜襟，无袖；窄身，直下摆，摆长及膝，两侧开衩，装饰简洁，线条流畅。 Standing collar, oblique front piece, sleeveless, narrow body part, straight hem reaching the knees, slits on both sides, with simple decorations and smooth lines.
20世纪40年代的短袖长旗袍 Short-sleeve full-length Chi-pao in the 1940s	立领，斜襟，短袖，仅包裹住肩部；衣身紧而窄，较为适体，衣摆长及脚踝，两侧开衩至膝部。 Standing collar, oblique front piece, short sleeves, covering only the shoulders; narrow body part fitting closely to the body, hem reaching the ankles, slits on both sides reaching the knees.
20世纪50年代的短袖旗袍 Short-sleeve Chi-pao in the 1950s	立领，斜襟，短袖；收腰，圆下摆，两侧开衩，长及小腿；胸前和衣摆处有简洁的花形装饰。 Standing collar, oblique front piece, short sleeves, contracted waist, round hem, slits on both sides, skirt reaching the shanks, simple flowery decorations on the chest and hem.
20世纪80年代的短袖旗袍 Short-sleeve Chi-pao in the 1980s	也称"花瓶式旗袍"。立领，斜襟，凸显腰身曲线，衣摆至膝部，两侧高开衩。古典与新潮结合。 Also called vase-shaped Chi-pao, standing collar, oblique front piece, lovely curves on body and waist parts, hem reaching the knees, high slits on both sides, combination of classical and fashionable elements.

旗袍之美
Beauty of Chi-pao

旗袍美得动人，美得文雅，美得知性，美得梦幻……以简洁、明朗的线条勾勒出了东方女性优美的曲线，以轻盈、细腻的面料映衬出了女人的优雅和温柔，以华美、绚丽的色彩渲染出了女人的柔美和温润，以细致、精良的做工展示了女人的精致和娴雅。

一件旗袍的诞生过程是十分繁复的，要经过周密的量体、巧妙的设计、精心的剪裁和缝制，以及镶、滚、嵌、盘、绣等装饰工艺。一件旗袍最终呈现在人们眼前时，每一个细节都透着精细。

The Chi-pao has elegant, reasonable, dreamlike and soul-touching beauty. It outlines the marvelous body curves of oriental women with simple and clear lines. It showcases their elegance and tenderness with light and refined materials. It displays their subtle beauty and mild smoothness with gorgeous colors. It demonstrates their delicacy and gracefulness with the first-class tailoring.

The making of a Chi-pao is a complicated process that includes accurate measurement, ingenious design, well-planned tailoring and stitching, and diversified decoration techniques such as lacing, edging, embedding, coiling and embroidering. Every detail of a Chi-pao reveals the painstaking work of its makers.

> 形之美

一件旗袍的形体之美主要体现在领、襟、肩、袖、胸、腰、臀、衣摆、开衩等细节。小巧而精致的立领可以衬托修长的脖颈，给人一种娉婷的美；宽窄适中的衣袖使肩膀显得柔美；收腰、收臀的设计可以凸显出女人丰满的胸部、柔软的细腰和饱满的臀部。

> Beauty of the Shape

The beauty of the shape of a Chi-pao can be seen in many details, including the collar, front piece, shoulder, sleeves, chest, waist, hip, hem and slit. The small and exquisite standing collar accentuates the slender neck of the wearer, giving her a slim and graceful beauty. The sleeves in the right size make the shoulders light and beautiful. The contracted waist and hip lines highlight women's plump breast, soft and thin waist, and full hips.

- 油画《春怨》（作者：刘文进）
  旗袍是内敛的，可以裹住女性的丰臀和细腰，从领口一直遮蔽到踝骨，纹丝不露；旗袍也是张扬的，一条衩从脚踝开到腿部。一张一弛增添了无限空间，她不轻佻、不拘谨，含蓄而优雅。

  The Canvas *a Lonely Beauty in Spring* (by Liu Wenjin)
  Chi-pao is close. It covers the wearer's plump yet slender body from neck to the ankles. Chi-pao is also wild. Its slit goes from the ankle to the leg. Such closeness and openness add boundless space to Chi-pao and make it neither frivolous nor rigid. We have to admit that Chi-pao is reserved and elegant.

领：严丝合缝的立领，领头平整而圆顺，呈对称状，使衣襟平滑而顺美，使女性的脖颈更显修长。

Collar: The standing collar goes neatly around the neck. The collarband is even, smooth, and symmetric. Its well-distributed design makes the dress smooth and beautiful. Meanwhile, the wearer's neck looks more slender.

肩：肩部秀美、柔和。

Shoulder: The shoulder part is beautiful and gentle.

袖：衣袖紧、窄、合体。

Sleeves: The sleeves of this Chi-pao have a proper size.

胸省：胸部饱满，收胸自然，将女性美展现得更加充分。

Chest Dart: With natural chest dart, this Chi-pao shows a full breast, which showcases women's beauty more fully.

腰省：腰部线条流畅，收腰自然，使腰部显得柔软、平坦、纤细。

Waist Dart: With natural waist dart, this Chi-pao has smooth lines at waist. As a result, the waist of the wearer will look firm, even and slender.

盘扣：盘扣平整、对称而牢固，均匀而饱满，位置适中。

Frog: The frog is even, neat, symmetric, firm, full and well-positioned.

臀形：收臀自然，丰满而适中的臀部可以很好地突出女性身材的曲线美。

Hip: This Chi-pao has natural contraction at hips. The moderately full hips can effectively highlight the curvaceous beauty of female figures.

开衩：开衩至膝部上方，合缝严密。

Slit: The slit goes above the knees. Its end is tightly stitched.

衣摆：衣摆顺直，长及足面，可以衬托出女性修长的身材。

Hem: The hem drops straight down to cover the feet. It helps display the wearer's slender figure.

滚边：滚边饱满而平滑，宽窄均匀。

Edging: The edging is full and even, and has a proper width.

- 祥云纹锦缎长旗袍

女性的头、颈、肩、臂、胸、腰、臀、腿，以及手、足形成了众多曲线，这些线条又巧妙地构成一个整体，在旗袍的作用下相得益彰。

A Brocade Full-length Chi-pao with Auspicious Cloud Patterns

The head, neck, shoulders, arms, chest, waist, hips, legs, hands and feet of a woman form many curves. These lines seamlessly merge into an integral whole, which is best demonstrated by a Chi-pao.

旗袍的襟形

旗袍款式的变化主要是襟形、袖式、领形等的变化。旗袍上的襟指除去袖子，前面的那一片。中国传统袍服从商、周时期开始就以采用开襟形式为主。开襟即襟是分开的，纽扣在胸前的叫"对开襟"，纽扣在右侧的叫"右开襟"，纽扣在左侧的叫"左开襟"。常见的旗袍襟形有如意襟、圆襟、直襟、方襟、琵琶襟、斜襟、双襟等。

Shape of the Front Piece of Chi-pao

Change of Chi-pao style refers mainly to the change of the shape of its front piece, sleeves, and collar. Traditional Chinese gown-style dresses mainly adopted the open-front-piece style since the Shang Dynasty (1600 B.C. – 1046 B.C.) and the Zhou Dynasty (1046 B.C. – 256 B.C.). If the button is in the middle, it is called the Opposite Front Piece; if the button is on the right, it is called the Right Front Piece; if the button is on the left, it is called the Left Front Piece. Common shapes of the front piece of Chi-pao include *ruyi* (an S-shaped ornamental object symbolizing good luck) shape, round shape, straight shape, square shape, *pipa* (a four-stringed Chinese lute) shape, oblique shape, and two-side shape.

- 如意襟旗袍

  "如意"是中国一种象征祥瑞的器物，头部为灵芝形或云形，柄微曲，供人玩赏。如意襟旗袍就是通过镶滚的方式把"如意"装点在旗袍上，"如意"可被放在肩头，也可被顺着衣襟做，还有的旗袍的裙摆开衩位置被镶滚了如意云头。

  A Chi-pao with a *Ruyi*-shaped Front Piece

  Ruyi is a Chinese ornamental object symbolizing good luck. With a head shaped like a ganoderma or a piece of cloud and an S-shaped handle, *ruyi* serves as an article for admiration. This Chi-pao has a *ruyi*-shaped lace on it. The *ruyi* pattern can be placed on the shoulder, along the front piece, or around the hem slit.

- **大圆襟旗袍**

  圆襟是旗袍常见的开襟方式，线条圆顺而流畅。

  A Chi-pao with a Large Round Front Piece

  In a round smooth shape, the round front piece is the commonest style for Chi-pao.

- **方襟旗袍**

  方襟方中带圆，显得含蓄而内敛，又富于变化。这种款式的旗袍适合不同脸型的女性穿着。

  A Chi-pao with a Square Front Piece

  The square front piece contains round shapes. It is reserved and full of changes, suitable for women of various face features.

- **直襟旗袍**

  直襟旗袍会使女性的身材显得修长，适合圆脸、身材丰满的女性穿着。直襟旗袍的一排盘扣具有很强的装饰性。

  A Chi-pao with a Straight Front Piece

  The Chi-pao with a straight front piece will make the wearer look more slender. It is suitable for women with a round face and a plump figure. The straight row of frogs has a strong decoration effect.

旗袍
Chi-pao

• 双襟旗袍
双襟给人均衡、对称的美感，显得端庄而大方，适合中老年女性穿着。

A Chi-pao with a Two-side Front Piece

The two-side front piece gives a balanced and symmetric sense of beauty. It is graceful and magnanimous and suitable for middle-aged and elderly women.

• 曲襟
曲襟的形状像带棱角的"S"，开口较大，容易穿着。

A Curved Front Piece

The curved front piece looks like an S with edges and corners. With a large opening, it is easy to put on.

• 斜襟旗袍
斜襟是从领口斜划过胸前的衣襟款式，穿起来具有古典韵味。

A Chi-pao with an Oblique Front Piece

The oblique front piece runs from collar to chest. It guarantees a classical taste.

- **中长襟**

  这种形式的襟表现为从领口斜划出一个不很明显的弧形，避开胸部，一直延伸至腰部；身侧配以一排不对称的花扣，作为装饰。

  A Medium and Long Front Piece

  The medium and long front piece shows an obscure arc comes down from the collarband. Avoiding the chest, it extends all the way to the waist. A row of asymmetric frogs is arranged on the body side as decoration.

- **双圆襟**

  双圆襟可突出俏丽多姿之感，不同于大圆襟的规矩而稳重。

  A Double-circle Front Piece

  The double-circle front piece is beautiful and varied in shapes. It is different from the mature and prudent large round front piece.

- **琵琶襟**

  大襟只掩至胸前，不到腋下。

  A Pipa-shaped Front Piece

  The large front piece only covers the chest and does not reach the armpit.

旗袍的领形

发展到现在，旗袍的领形十分多样，常见的领形有高领、低领、无领、元宝领、波浪领、水滴领、V字领等。为了保证旗袍的领是硬挺的，传统的裁缝一般会用浆糊将白布浆硬，放入领内；对于有些由高级面料制成的旗袍，低于领口处通常被手工缝上一条刮浆白棉布，便于拆洗。

Shape of the Collar of Chi-pao

Today, Chi-pao has diversified collar shapes, including high collar, low collar, shoe-shaped collar, wave-shaped collar, water-drop-shaped collar and V-shaped collar. There is also the Chi-pao without a collar. To make the collar rigid, a traditional way used by a tailor is to harden a piece of white cloth by starching it with paste. Then, the cloth is put into the collar. To facilitate washing, a piece of starched white cotton cloth is usually stitched manually onto the Chi-pao made of first-class materials at a position lower than the collarband.

• 现代旗袍立领的典型款式
The Typical Standing-collar Design of Modern Chi-pao

旗袍的各种领形
Various Collar Shapes of Chi-pao

- 元宝领

元宝领斜压在下巴两侧，起修饰脸型的作用。着带有元宝领的旗袍时，需要抬高下巴、挺直脖颈，才能显出仪态的端庄。

A Shoe-shaped Collar

The shoe-shaped collar sticks to the two sides of the neck under the chins. It adorns the facial features. Wearing such Chi-pao, one must lift her chin and straighten up her neck to present a dignified bearing.

- 上海领

上海领适合各种脸型的女性，此领型的旗袍能使穿着者显得谦和而柔美。

Shanghai Collar

Shanghai collar is suitable for women of various facial features. This kind of collar can make the wearer look humble and beautiful.

- 波浪领

波浪领的旗袍风格活泼，适合年轻女性穿着。

Wave-shaped Collar

The wave-shaped collar has a vivacious style and is suitable for young ladies.

- 方领

方领方中有圆，风格庄重而严谨。

Square Collar

The square collar contains round lines and is solemn and upright in style.

- 水滴领

水滴领表现为在领口处挖出水滴的形状，露出些许肌肤，别具风情。

Water-drop-shaped Collar

The water-drop-shaped collar exposes some area under the neck and shows unique charm.

- V字领

V字领旗袍适合肩比较宽的女性穿着。

V-shaped Collar

The V-shaped collar is suitable for women with relatively wide shoulders.

旗袍的袖形

旗袍的袖型常随潮流而变化。时而流行长袖，长过手腕；时而流行短袖，露肘。旗袍的袖子在装饰上也求新、求异，常见的袖形有长袖、短袖、开衩袖、荷叶袖、喇叭袖等，有些旗袍无袖。

Shape of the Sleeves of Chi-pao

The sleeves of Chi-pao often change with the trend. Sometimes the long sleeves are in fashion and they reach beyond the wrists. At other times, however, the short sleeves are popular and they can hardly reach the elbows. The decorations on the sleeves of Chi-pao are also diversified and never cease to change. The common sleeves include long sleeves, short sleeves sleeves with slits, lotus-shaped sleeves and horn-shaped sleeves. Some Chi-pao has no sleeves at all.

- 清代早期旗袍的箭袖

箭袖是身份的象征，代表着勇敢、坚韧和尊贵，清代的满族人常以箭袖旗袍为礼服。

An Arrow Sleeve on the Chi-pao in the Early Period of the Qing Dynasty (1616–1911)

Representing bravery, tenacity and dignity, arrow sleeves are the symbol of the wearer's high social status. The Manchu people in the Qing Dynasty (1616-1911) often wore the Chi-pao with arrow sleeves as their formal attire.

- 现代旗袍的典型短袖

A Typical Short Sleeve on Modern Chi-pao

各种样式的传统旗袍袖
Various Traditional Chi-pao Sleeves

- 镶蕾丝边的中袖
A Middle-length Sleeve with Silk Lacing

- 镶线边的喇叭袖
A Horn-shaped Sleeve with Line Lacing

- 紧、窄、合体的长袖
A Long Sleeve in Proper Width

- 镶线边的中袖
A Middle-length Sleeve with Line Lacing

- 垂至腕部的荷叶袖
A Lotus-shaped Sleeve Dropping to the Wrist

- 镶蕾丝边的开衩袖
A Sleeve with Silk Laced Slit

- 镶锯齿形边的喇叭袖
A Horn-shaped Sleeve with Sawtooth Lacing

- 简洁而清爽的短袖
A Simple and Refreshing Short Sleeve

- 镶线边、蕾丝边的喇叭袖
A Horn-shaped Sleeve with Line and Silk Lacing

旗袍的裙摆

传统的旗袍裙比较宽松；后来随着腰身的收拢逐渐变短，特别是两边的开衩，或大或小，行走时下角微微飘动，给人以轻快、活泼之感。发展到现在，旗袍的裙摆有宽摆、直摆、A字摆、礼服摆、鱼尾摆、前短后长、锯齿摆等形式。

Hem of Chi-pao

Traditional Chi-pao skirt was loose. Later, with the waist being tightened up, it gradually became short. The slits on both sides, big or small, slightly flutter when the wearer is walking, giving her a breezy and vivacious look. Now, Chi-pao has many kinds of hems, including wide hem, straight hem, A-shaped hem, formal attire hem, fish tail hem, short-front and long-rear hem, and sawtooth hem.

- **身穿马甲旗袍的女子（20世纪20年代）**

  20世纪20年代，上海流行马甲旗袍，被罩在短袄外。后来，长马甲与小短袄合成一件，长至腿部，大袖口，造型仍是直线型，腰线较低，曲线不明显，下摆至膝。

  A Woman in Vest Chi-pao (1920s)

  In the 1920s, the vest Chi-pao was popular in Shanghai. The vest was worn over the short coat. Later, a long vest and the short coat merge into one piece, which reaches the leg. Vest Chi-pao has large cuffs and a straight line design. Its waist line is relatively low without an obvious curve. The hem reaches the knees.

这位女子采用的是传统装扮：梳盘发髻，留刘海；身穿一件低领、大袖旗袍，裙摆长至脚踝，衣领、袖口、裙摆都带有繁复的滚边装饰；脚上穿着一双绣花鞋，鞋面的绣花与衣身的滚边相互映衬。

This woman is in a traditional attire. Her hair is in a bun and she has bangs on her forehead. She wears a Chi-pao with a low collar and large sleeves. The hem drops to her ankles. The collarband, cuffs and hem have complex edging decorations. She wears a pair of embroidered shoes. The embroidery on the shoes and the edging on the Chi-pao set each other off beautifully.

这位女子留着齐耳短发；旗袍的领子略高，裙摆略短，衣缘的滚边装饰简约、素雅；脚上穿着一双高跟鞋。

This woman has short hair aligning with her ears. Her Chi-pao has a relatively high collar and a short hem. The edging on the Chi-pao is simple and elegant. She wears a pair of high-heeled shoes.

这位女子的发型与前一位没有太大变化；旗袍袖子适中，裙摆更短，露出了小腿，衣缘滚边上装饰着抽象的条纹，线条简洁而明快。

This woman has almost the same hairdo as the previous one. Her sleeves have a moderate width and her hem is shorter, which exposes her shanks. The edging consists of some abstract yet simple stripes.

这位女子留着男式短发，显得英姿飒爽；穿着暗扣旗袍，袖子较短，露出了肘部，裙摆也提至膝盖处。通过画面上简笔勾勒的阴影，可以看出，旗袍更加合身，女性动人的曲线被更加明显地展露出来。

This woman has a man's short hairdo, which gives her an elegant and unconventional look. The frogs on her Chi-pao are all covert ones. The sleeves are short to expose her elbows and the hem is also lifted to her knees. From the shadowy sketches in the picture, we can see that Chi-pao fits in well with her figure. Her lovely body lines are brilliantly shown.

- **20世纪20年代的各种旗袍款式**
  这是1927年《上海画报》中的一幅插图，它像一面镜子一样反映了当时的旗袍面貌。
  The Various Chi-pao Designs in the 1920s
  It was an illustration in the *Shanghai Pictorial* in 1927. It mirrors the Chi-pao styles at that time.

这位女子穿着宽身旗袍，宽松的袍身却抵不住青春气息的自然流露；袖口采用双层滚边，亮丽的红色花边在黑白搭配中很醒目。

Though in a loose Chi-pao, this woman still has around her a strong sense of youth. The cuffs have double-layer edging. The bright red lace is an eye-catching decoration against the black-and-white background.

这位女子穿着马甲旗袍，衣领、裙摆、袖口都形如锯齿，带有夸张的花边装饰。这种款式显然受西式服装风格影响而产生。

This woman wears a vest Chi-pao. Its collar, hem and cuffs all have exaggerated sawtooth lacing. This style is obviously the result of the influence of Western costumes.

- **穿着大袖旗袍的女人（20世纪30年代）**

  20世纪30年代，来自欧美的高跟鞋开始被用来与旗袍搭配，改变了旗袍的流行风尚。此时的旗袍裙摆渐渐变长。精巧的高跟鞋更增加了旗袍的美感，使女性身材更显纤细、修长。随着旗袍下摆的加长，开衩也越来越高。

  Women in the Chi-pao with Large Sleeves (1930s)

  In the 1930s, the high-heeled shoes from Europe and America began going along with Chi-pao. This changed the fashion trend of Chi-pao. Its hem gradually went down and the exquisite high-heeled shoes added beauty to Chi-pao. Together, they made women slenderer. With the hem going down, the slit went up higher and higher.

- **现代旗袍的裙摆**

  如今，旗袍以多变的姿态展现着女性美，演绎着别样的东方风情。

  Hem of Modern Chi-pao

  Now, Chi-pao is displaying feminine beauty and showcasing the unique oriental charm in diversified styles.

> 工艺之美

旗袍具有很多细致而富有特色的制作工艺，主要包括镶（镶边）、滚（滚边）、嵌（嵌条）、荡（荡条）、盘（盘扣）、绣（刺绣）等。这些精巧、繁复的工艺使旗袍在细节上环环相配，整体上更加精致、美观。一件旗袍的诞生，

> Beauty of the Make

Chi-pao is made through many refined and distinct tailoring procedures, mainly including lacing, edging, embedding (panel), swinging (swinging piece), coiling (frog) and embroidering. These exquisite and complicated techniques cooperate with each other in the details of a Chi-pao and make it more delicate and beautiful as a whole. Several techniques will be used in the making of a Chi-pao. For example,

- 用来制作旗袍的缝纫机

缝纫机是20世纪初才开始在中国逐渐流行起来的，并被应用到旗袍制作中。在应用缝纫机之前，传统旗袍的制作工具有十余种，其中大部分已渐渐远离现代人的生活；而品味这些原汁原味的传统手工艺，正是了解中国旗袍文化的重要一步。

A Sewing Machine for Making Chi-pao

The Sewing machine was not popular in China until the early 20th century. It was then used in making Chi-pao. Before that, over ten kinds of tools were needed for making a traditional Chi-pao. Now, most of these tools are nowhere to be found. However, reviewing and appreciating these traditional handicrafts happens to be the major step to understand Chinese Chi-pao culture.

几种工艺通常被搭配使用，或在局部进行刺绣，或做个样式别致的盘扣，轻巧而雅致。

some Chi-pao has embroidery and some others have unconventional frogs, which may be light and graceful.

• 缝盘扣

制作传统旗袍与制作传统中式服装使用的工具相同，一把尺子、一把剪刀、一个熨斗、几枚针就是裁缝的全部工具。经验丰富的裁缝先用尺子和剪刀裁剪出旗袍的精妙轮廓，再一针一线细致地缝合衣片，最后用熨斗归拔出每个部位的细节。

Stitching of Frogs

A traditional Chi-pao is made with the same tools for making traditional Chinese clothes. All a tailor needs is a ruler, a pair of scissors, an iron and several needles. The experienced tailor will use the ruler and scissors first to cut out the exact outline of the Chi-pao. Then, he will finely stitch the pieces together. Finally, he will use the iron to block into shape the details of every part.

归拔

为了恰到好处地体现出女性的曲线美，归拔工艺常被用来处理细节，以凸显胸部、腰部、臀部等处的线条，使平面裁剪的旗袍产生立体的效果，甚至不用穿在身上就能体现出人体造型的韵味。"归"是将衣片上需归拔的部位向内侧缓慢推进，使衣片归拔部位的边长变短，归拔部位形成隆起的形状。"拔"是将衣片需拔开的部位向外侧拉开，使衣片拔烫部位的边长增长，以适应曲线变化大的部位。

Blocking

To better display feminine beauty by accentuating lines of women's breast, hips and waist, the blocking process is needed to deal with some details. In this way, the Chi-pao can have a 3D tailoring effect even though it only goes through plane tailoring. In fact, a Chi-pao can demonstrate the charm of the human figure on its own. The blocking process contains two techniques, namely pressing in and pressing out. Pressing in is to slowly press inwards the parts on the cloth to be collected. It shortens the side length of these parts and makes them into a bulge shape. Pressing out is to press outwards the parts on the cloth to be spread out. It elongates the side length of these parts and adapts them to other parts with large curve changes.

传统旗袍的制作工具
Tools for Making a Traditional Chi-pao

缝衣针 Needle	传统旗袍是裁缝全手工一针一线缝制而成的，因此缝衣针是非常重要的工具。手工缝制旗袍，细节的处理全凭裁缝的手感和经验，要求缝制间距平均、松紧适度。 Traditional Chi-pao was made stitch by stitch by tailors. Therefore, the needle was the very important tools. To make Chi-pao manually, the tailor totally relied on his touch and experience. The spacing between stitches should be even and the finished Chi-pao should be neither tight nor loose.
顶针 Thimble	顶针形同圆环，表面均匀分布着凹窝。顶针一般被套在右手中指上，用来保护手指不被针刺伤。顶针多由银、铜、铅或其他金属制成。 A thimble is a ring with evenly distributed pits on its surface. A thimble is normally worn on the middle finger of the right hand to protect the finger from being injured by the needle. In the past, thimbles were mainly made of silver, copper, lead, or other metals.
尺子 Ruler	在制作传统旗袍的过程中，裁缝使用旧制市尺，通常分直尺和软尺两种。直尺用于制图打版和裁剪时进行测量和画线，软尺则用于量体。 In making the traditional Chi-pao, a tailor used *chi* as the measurement unit. There were two kinds of rulers: straight scale and band tape. The straight scale was for measuring and drawing lines during drafting and tailoring while the band tape was for body measuring.
划粉 Tailor's Chalk	划粉是一种裁剪之前在衣料上画线、定位的辅助用具，多以各种颜色的石粉制成薄片，确保裁剪的衣料精确定版。 The chalk is an auxiliary tool for drawing lines and positioning the fabric materials before tailoring. It is normally a thin flake made of mountain flour in various colors. It is used to guarantee the precise cutting of the materials.
针包 Needle Kit	针包多以有弹性的棉花或毛线塞制而成，用来插放、收集经常使用的缝衣针、大头针，这样能使针体保持滑润，不生锈。针包上缝有松紧带，可戴在手腕上。 The needle kit is normally made of springy cotton or knitting wool. It is for collecting the needles and pins often used in tailoring. It keeps the needles smooth and free of rust. There is an elastic band on the kit, so that the kit can be worn on the wrist.

刮浆刀 Paste Blade	制作传统旗袍，很多工艺环节都需要刮浆刀，在面料上刮糨糊，如嵌条、做盘扣、制领子、裁剪等；特别是做真丝旗袍时，真丝面料质地光滑，裁剪时容易跑位，用刮浆刀可保证面料位置不动。 In making a traditional Chi-pao, a paste blade was needed in many procedures. It was used to smear paste over cloth when making panels, frogs and collars, and during tailoring. It was especially indispensable when making a silk Chi-pao. Silk is so smooth that it moves easily. A tailor often used a paste blade to keep the material in position.
绣花剪刀 Embroidering Scissors	绣花剪刀是裁缝绣花时必用的工具，可以用来修剪线头。 The embroidering scissors must be used during embroidering. It is for trimming the end of threads.
剪刀 Scissors	剪刀是制作传统旗袍最重要的工具。裁布料时多一分、少一分都不行，否则制作出来的旗袍效果会相差甚远。裁剪面料时，多用手感舒适、刃口锋利、开合顺利的剪刀。中国的王麻子、张小泉牌剪刀质地精良，都是享誉国内外的著名品牌。 A pair of scissors was an important tool in traditional Chi-pao making. Cloth cutting should be precise; otherwise the finished Chi-pao would be substandard. During cloth cutting, scissors with moderate touch, sharp blades, and smooth open and close movements were used. The Wang Mazi and Zhang Xiaoquan brand scissors of China are high-quality products well-known worldwide.
锥子 Stiletto	锥子用来处理领角、衣角等细节，使旗袍的边角部分干净利落，不毛糙。 A stiletto was for handling details such as collar corner and hem corner. The finished Chi-pao should have neat and smooth corners.
镊子 Tweezers	镊子是制作旗袍的辅助工具，用来调整线条、控制缝料的松紧度。做盘扣时，镊子常被用来处理细节。 A pair of tweezers was an auxiliary tool for making a Chi-pao. It was used to adjust lines and control the tightness of the materials. When making frogs, a pair of tweezers was often used to handle the details.
熨斗 Iron	熨斗最初为铜制，后改为铁制，即把烧红的木炭放在熨斗里，将底部烧烫后用来熨烫旗袍。裁缝通常在熨斗和旗袍之间放一块烫布，使旗袍面料不被烫坏。 The iron was initially made of copper. Later, it was made of iron. People used to put burning charcoal into an iron and used it to press out the Chi-pao. The tailor usually put a piece of cloth between the iron and the Chi-pao to protect the Chi-pao material from being damaged.
喷壶 Sprinkling Can	制作旗袍时，一些面料局部需要被进行归拔处理，可用喷壶喷湿特定部位，然后再进行熨烫。 In making a Chi-pao, some parts of the materials require a handling process called blocking. In the process, the sprinkling can is used to moisten the specific parts before ironing.

镶边

　　镶边是非常独特的旗袍装饰工艺，即用不同颜色的布料对旗袍的衣襟、领口、袖口、开衩、底边等边缘部位进行缝合拼接，布料上多带有各色、各样的绣花。镶边用的布料一般以暗针直接缝在服装的表面。

　　镶边工艺在清代非常流行，无论是满族女子还是汉族女子，都喜欢在服装上镶一道道花边，有"三镶""五镶"，最多可至"十八镶"。19世纪末，各种机织花边装饰成为新的旗袍装饰手法。如今，旗袍的风格从繁复趋于简洁，镶边变得细而窄，很多旗袍甚至无镶边。

Lacing

Lacing is a unique decoration technique in making a Chi-pao. Cloth pieces in different colors are stitched to the edge of the front piece, collar, cuff, slit and hem of the Chi-pao. On the cloth are diversified embroideries. Normally, lacing is directly stitched to the surface of a Chi-pao with blind stitches.

　　Lacing technique was very popular in the Qing Dynasty (1616-1911). Both Manchu and Han girls liked to have lacing on their clothes. There were three laces, five laces, and at most eighteen laces. By the end of the 19th century, various machine-made laces became new decorations for Chi-pao. Now, Chi-pao has changed from complicated to simple. The present-day Chi-pao has narrow lacing or no lacing at all.

- **现代旗袍的镶边工艺**
  经过面料改造，普通的小花棉布被在边缘处进行了刺绣处理。白色的蕾丝使棉布具有了新生命。穿着这款旗袍会给人以亲切感。

  Lacing on Modern Chi-pao
  Going through material transformation, the ordinary floret cotton cloth receives embroidery on its edge, which, seen as white lace, gives a new lease of life to the cotton cloth. This Chi-pao guarantees a friendly atmosphere around the wearer.

领：小立领起保暖作用，领口镶以黑、蓝两道花边。

Collar: The small standing collar helps to keep warmth. The collarband has two laces, one black and one blue.

襟：衣襟上镶有双层花边，里层是花卉纹窄花边，外层是"蝶恋花"纹宽花边。

Front Piece: The front piece has double-layer lacing. The inner layer is a narrow flowery lace while the outer layer is a wide lace with the "Butterfly Loving the Flower" pattern.

开衩：衣身两侧开衩，顶端镶有"如意云头"纹。

Slit: There are slits on both sides. On top of the slits are the "Auspicious Cloud" patterns.

袖：衣袖宽而直，绣有精美的"蝶恋花"纹。

Sleeves: The sleeves are loose and straight, with the exquisite pattern known as "Butterfly Loving the Flower".

衣摆：衣摆略宽，侧边和底边也都镶有双层花边，样式与衣襟的镶边相同。

Hem: The hem is a little bit wide. Both its bottom and two sides have the same double-layer lacing as that on the front piece.

- 清代暗纹绸缎镶边旗袍

这是清代早期的旗袍，做工精细，用料考究，衣身不加任何装饰，与精致、华美的人镶边形成鲜明的对比。

A Chi-pao with Dark Pattern Brocade Lacing in the Qing Dynasty (1616-1911)

This Chi-pao was made in the early period of the Qing Dynasty. It shows refined tailoring and exquisite materials. Its main part has no decoration, which forms an interesting contrast with the large elegant and gorgeous lacing.

滚边

　　滚边工艺是用来包裹旗袍的开衩、领口、袖口、底边等开口部位的。滚边工艺非常考究，常与镶边工艺配合使用。滚边工艺最难把握之处在于在衣服转弯处保持线条流畅。制作用来滚边的滚条时，可以选择与旗袍底色接近的布料，或是选择与旗袍上的图案颜色相近的布料。最为常见的滚边为丝质滚边，外观圆润、硬挺。

Edging

Edging technique is for covering the openings at the slits, collarband, cuffs and hem of the Chi-pao. A refined process, it is often used together with the lacing technique. The most difficult task of edging is to keep the line at the turn of the clothes smooth. The edging can be made of the cloth in a color similar to the background color of the Chi-pao or with the pattern similar to those on the Chi-pao. The commonest edging is silk edging with a round and hard appearance.

- 滚边
  Edging

领部的滚边：衣领处被镶滚了一道红边，与旗袍的花色相符。衣领处花团锦簇，有盛开的大朵牡丹，有含苞待放的花蕾，还有嫩绿的枝叶，非常引人注目。

Edging on the collar: The collar has a red lace, which is consistent with the flowery pattern of the Chi-pao. The collar easily attracts attention due to the concentrated flowery patterns, including the large blossoming peony, the full buds and the tender green branches.

袖口的滚边：袖口、衣襟处都被镶滚了一道红边，金色的纽扣与雍容华贵的牡丹相互呼应。

Edging on the cuffs: Both the cuffs and front piece have a red lace. The golden buttons work in concert with the dignified and graceful peonies.

- 国画牡丹旗袍

这件旗袍融合了中国国画和古典服装的精髓，娇艳的红、白、粉色牡丹遍布衣身，花形饱满，花色艳丽，与质地光滑、细腻的白色丝绸底料相得益彰。

A Chi-pao with Peony Pattern Drawn in Traditional Chinese Painting Technique

This Chi-pao combines the essence of traditional Chinese painting and classical clothing. Delicate and charming, the red, white and pink peonies cover the entire Chi-pao. They are in full blossom and going well with the fine and smooth white silk material.

嵌条

由于旗袍曲线变化大，为保证侧缝曲线的稳定性，需要在旗袍一些部位的侧缝开口处沿着缝头贴嵌条。嵌条也是一种非常重要的旗袍装饰工艺，即将约1厘米宽的布条缝在两块衣片的边缝之间。嵌条工艺能增强旗袍的立体感，常与镶边工艺搭配使用。在缝合旗袍的衣片时，嵌条起着美化衣服的作用，尤其是旗袍的肩、腰、臀等特定部分的立体造型，都离不开嵌条工艺。

Panel

Chi-pao has large curve changes. To guarantee the stability of its side curves, some parts of the Chi-pao need a panel pasted along the sewing of the side opening. Panel is another very important decoration for Chi-pao. It is in fact a one-centimeter-wide cloth strip stitched between two pieces of cloth. Panel can enhance the 3D effect of the Chi-pao and is often used together with lacing technique. When the cloth strips of the Chi-pao are stitched together, panel plays a beautifying role. Panel is indispensable especially in the 3D modeling of some specific parts of the Chi-pao, including the shoulder, waist and hip.

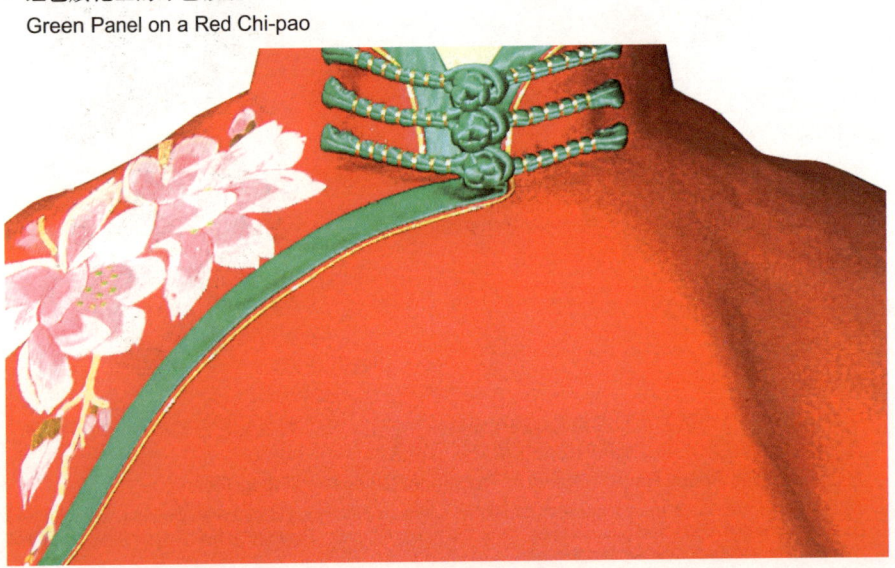

- 红色旗袍上的绿色嵌条
  Green Panel on a Red Chi-pao

荡条

荡条工艺也是一种条状的装饰形式，一般选用不同于旗袍材质的面料缝在旗袍的领口、袖口、腰身、底边等部位。

Swinging Piece

Swinging piece is a strip decoration. Made of a material different from that of the Chi-pao, such strip is normally stitched to the collar, cuff, waist and hem of the Chi-pao.

- **现代旗袍精致的荡条装饰**

  事实上，这件现代旗袍腰部的装饰早已突破了荡条的范畴，体现了现代款与传统样式的完美结合。

  Exquisite Swinging Piece on a Modern Chi-pao

  Actually, the waist decoration on this modern Chi-pao does not belong to the traditional swinging piece, and reflects a perfect combination of modern and traditional style.

省道

　　省道就是在缝制旗袍前、后衣片时，去除肩部、胸部、腰部、臀部等部位的多余衣料，使面料更加贴合人体曲线变化，从而让旗袍更加合身、适体。早期的旗袍制作没有省道工艺，线条平直，衣身宽松，胸、腰围度与衣裙的尺寸比例较接近。20世纪30年代中后期，旗袍的制作引入了西方工艺，衣片上出现了省道，使得旗袍更加合体，旗袍的曲线美得以完美地呈现，产生了立体效果。

Dart

Dart is the surplus cloth removed from some parts of the Chi-pao, including the shoulder, chest, waist and hip, before its front and back pieces are stitched together. By doing so, the material will follow the curve of human body more smoothly and the Chi-pao will fit better. During the making of the earliest Chi-pao, there was no dart. The Chi-pao had only straight lines and loose body parts. The width at the chest and waist was close to the size of the dress. In the middle and late 1930s, Western techniques were introduced to the making of Chi-pao. With dart removed, the waist fitted better to the body. The line of beauty of the Chi-pao was perfectly demonstrated. There was also a 3D effect.

- **省道工艺使旗袍更合体**
  这款旗袍采用西瓜红色面料，看上去很清新。胸前的手绘玉兰花活灵活现，花心处的点钻处理增加了华丽感，非常适合在结婚时或晚会活动上穿着。同时，这款旗袍的省道工艺非常精湛，具有塑形的效果。

  Dart Technique Making the Chi-pao Fit Better
  This Chi-pao is made of a water-melon red material. Looking fresh, it has a vivid hand-painted yulan magnolia on the chest. The floral spots in the center add resplendence to the pattern, making it an ideal dress for weddings or evening parties. Meanwhile, the dart technique is used skillfully to give this Chi-pao the best shaping power.

旗袍的盘扣

旗袍的盘扣也是一大特色工艺，又称"盘纽"，是用布料细条编织而成的。盘扣不仅具有固定衣襟的功能，更能展现独特的美感。盘扣的制作工艺和款式造型是我国传统服装的独特标志之一，用于旗袍的领口、衣襟、开衩等部位。

除了布扣，盘扣还有金、银、铜、翡翠、琉璃、宝石等材质。在制作直扣、琵琶扣、蝴蝶扣及实芯花扣时，讲究造型饱满、做工细致。若用薄面料制作盘扣，需要衬几根棉纱线或丝绳，起使盘扣坚固的作用；若用厚面料，可以不用绳线衬托。制作造型复杂的盘扣，可以使用包铜丝法：先裁剪出斜布条，将布条两端的毛口向内折起，用一根细铜丝夹在中间，用镊子折叠出各种造型。

Frogs of Chi-pao

Frog is another major signature technique used in making a Chi-pao. Also known as the frog button, it is woven with cloth strips. Frog can not only fix the front piece, but also serve as a unique decoration. The frog-making technique and its style are one of the unique symbols of traditional Chinese costumes, and it is used at its collarband, front piece and slit.

Besides cloth, frogs can also be made of gold, silver, copper, emerald, azure stone and gem. In making the straight button, *pipa*-shaped button, butterfly button and solid core flower button, the goal is to make the button full in shape and exquisite in workmanship. If thin materials are used to make the frog, several pieces of cotton yarn or silk band are needed to make the frog strong and sturdy. If thick materials are used, the said cotton yarn or silk band can be omitted. To make the frog with complicated and changing shapes, the copper cladding method can be adopted. First, cut out an oblique piece of cloth and fold up the burrs at two ends. Then, place in between a piece of thin copper wire and twist it into various shapes with a pair of tweezers.

盘扣造型繁多，植物造型的有梅花扣、菊花扣、玫瑰扣等，动物造型的有凤凰扣、孔雀扣、燕子扣、蝴蝶扣、蜻蜓扣等，还有仿中国结造型的，比如吉祥结扣、如意结扣、同心结扣等，以及仿汉字造型的，比如"一"字扣、"万"字扣、"吉"字扣、"喜"字扣等。

Frogs can be designed in various shapes. Some take plant shapes such as plum blossom, chrysanthemum, and rose. Some take animal shapes such as phoenix, peacock, swallow, butterfly and dragonfly. Still some take the shape of Chinese knots such as auspicious knot, *ruyi* knot and true lover's knot. There are also frogs shaped like Chinese characters, including 一 (one), 万 (ten thousand), 吉 (auspicious) and 喜 (happiness).

旗袍的刺绣工艺

　　刺绣工艺是中国优秀的传统手工艺之一，至今已有数千年的历史。中国的刺绣源远流长，在世界服饰文化宝库中占有重要的位置，是中华民族智慧的结晶。

　　刺绣就是在布帛、丝绸、锦缎等织物上用金、银线或丝、棉、绒

- 旗袍上的"彩蝶飞舞"刺绣

采用金、银丝线手绣出旗袍的蝴蝶图案，线条简单，富有生趣。

The Embroidery of "Dancing Butterfly" on a Chi-pao

This Chi-pao has some butterfly patterns embroidered with gold and silver threads. The simple lines form a vivid and interesting image.

Embroidery on Chi-pao

Having a history of several millenniums, embroidery is one of the excellent traditional handicrafts of China. It occupies an important position in the treasure-house of world clothing culture. It is the fruit of the wisdom of the Chinese nation.

　　Embroidery is the various decorative patterns stitched on fabrics such as cloth, silk and brocade with gold and silver threads or colored threads made of silk, cotton, velvet, etc. A thin steel needle goes up and down and creates all kinds of beautiful images, patterns or characters.

　　Embroidery can be seen all over China. Different places have their

- 刺绣半成品

刺绣时，先要将织物底料固定在绣架上。

A Semi-finished Embroidery

During embroidering, the base fabric should be fastened onto an embroidery stand.

● 京绣

京绣是以北京刺绣为中心的刺绣品的总称。京绣选料考究，针工巧妙，样式大方，图案丰富，色彩绚丽，格调高雅，最大的特点是绣线配色艳丽，色彩与瓷器中的粉彩、珐琅色有异曲同工之妙。此外，京绣做工严谨，一针一线都精致、华贵，严格遵循"图必有意，纹必吉祥"的宗旨。

Beijing Embroidery

Beijing embroidery is the general term for the embroidery works made mainly in Beijing. Beijing embroidery is known for its refined materials, ingenious needlework, tasteful design, diversified patterns, rich colors, and elegant style. Its biggest feature is the gorgeous coloration accomplished with the threads, which achieves the same result with the famille rose and enamel color of the chinaware. In addition, Beijing embroidery boasts exquisite workmanship. Every stitch demonstrates the theme of delicacy and luxury and strictly follows the principle that "every picture has a meaning and every pattern is auspicious".

等材质的彩线绣制各种装饰图案。仅凭借一根细小的钢针上下穿刺运动，就能制造出各种优美的图像、花纹或文字。

中国的刺绣工艺几乎遍布全国各地，不同地区的刺绣题材、用线、用色、针法等各具特色，其中

own embroidery themes, thread uses, colorations, and needle works. Among the most typical ones are Beijing Embroidery, Suzhou Embroidery, Hunan Embroidery, Sichuan Embroidery, and Guangdong Embroidery. Embroidery has been widely used in Chi-pao making since the Qing Dynasty (1616-1911). The

比较有代表性的刺绣有京绣、苏绣、湘绣、蜀绣、粤绣等。早在清代，刺绣工艺就已被广泛应用在旗袍的制作上了，精美的纹样使旗袍锦上添花。

exquisite patterns give an added grace to the already beautiful Chi-pao.

- **用于刺绣的彩色丝线**

在刺绣前，要先按照绣稿挑选绣线。为了增强表现力，表现一种色彩需要准备从深到浅十几个色级的绣线。根据绣稿和不同针法的需要，还要搭配粗细不同的色线。有的色线需要被劈成几丝。

Colorful Silk Threads for Embroidery

Before embroidering, threads shall be chosen according to the embroidery draft. To enhance the impression given by the pattern, a single color will be demonstrated with the embroidery threads at over a dozen color levels from deep to light. Based on the requirements set by the embroidery draft and different needleworks, the colorful threads in different thicknesses shall be used together. Some threads even have to be split up into several strands.

- **牡丹纹蜀绣**

蜀绣是以四川成都刺绣为代表的刺绣品的总称。蜀绣的原料以彩丝和软缎为主，常见题材有山水、人物、花鸟、鱼虫、飞禽、走兽。蜀绣绣工精细，针法严谨，色彩明快，图案美观，风格简洁，具有浓郁的地方特色。

Sichuan Embroidery in Peony Pattern

Sichuan embroidery is the general term for the embroidery works made mainly in Chengdu of Sichuan Province. Sichuan embroidery is worked out mainly with colorful silk and soft satin. Its common themes include landscapes, human figures, flowers, fish, insects, birds and beasts. Sichuan embroidery is known for its exquisite workmanship, rigorous needle work, bright colors, beautiful patterns, simple style and rich local characteristics.

刺绣工具

旗袍被缝制完毕，其局部有时会加以刺绣，需要借助绣框和绣架。绣框多为圆形，带有装钳夹、脚架，用来固定局部衣料。进行大面积的刺绣时，可以使用绣架，这样刺绣者就能腾出双手，便于操作。

Embroidery Tools

After a Chi-pao is stitched, sometimes embroidery will be added to some parts of it. To do so, an embroidery frame and stand are needed. The frame is normally a round one with a forceps holder and a pedestal for fastening parts of the Chi-pao. To make large-area embroidery, a stand can be used to free both hands of the embroiderer for easier operation.

- 绣框

An Embroidery Frame

- 荷花纹苏绣

苏绣是以苏州刺绣为代表的刺绣品的总称。苏绣工艺风格独特，绣工精湛，针法灵活而细腻，图案秀丽，用色雅致，形象传神。

Suzhou Embroidery in Lotus Pattern

Suzhou embroidery is the general term for the embroidery works made mainly in Suzhou. Suzhou embroidery is known for its unique handicraft style, exquisite workmanship, flexible and refined needlework, pretty patterns, elegant coloration and vivid images.

女红

中国古代的女人大都擅长做针线活，包括纺织、缝纫、刺绣等，称为"女红（gōng）"。中国封建社会一直存有"男耕女织"的传统思想，女子从小就要学习描花、刺绣、纺纱、织布、裁衣、缝纫等女红。女红在江南尤其受重视。作为中国传统文化的一部分，女红具有独特的魅力，而且与人们的日常生活密不可分，也与各民族、各区域的风俗紧密相连，与深厚的社会文化一脉相承。

Needlework

In ancient China, most women were good at needlework, including spinning, weaving, sewing and embroidering, and they are called "nǚ gōng". The Chinese feudal society had always kept the traditional idea of "men for farm work and women for spinning and weaving". Girls had to learn needlework at an early age, which included flower drawing, embroidering, spinning, weaving, dress making and sewing. This was especially the case in the area south of the Yangtze River. As part of traditional Chinese culture, needlework has a unique charm and is closely interwoven with people's daily life and the customs of different ethnic groups and different places. It is in the same strain with the profound social culture.

- 中国工艺美术大师蒋雪英正在刺绣

刺绣时，一只手在绣架上面，一只手在绣架底下，针刺出、刺入，循环往复，直到取得预期的效果。

Jiang Xueying, a Master in Chinese Arts and Crafts, is Embroidering

During embroidering, the embroiderer has one hand above the embroidery stand and the other hand below. The needle goes up and down repeatedly until the desired result is achieved.

• "锦鸡牡丹"纹湘绣

湘绣是以湖南长沙刺绣为代表的刺绣品的总称。湘绣绣工精湛，色彩鲜明，风格质朴；绣品上的绒面花纹富有质感，颜色浓淡相宜，形象生动、逼真。以中国画为题材是湘绣令人称道之处，绣工运用上百种颜色的绣线、数十种针法，融绘画、书法、刺绣为一体，构图美观，色彩丰富，绣品如画。

Hunan Embroidery in "Golden Pheasant and Peony" Pattern

Hunan embroidery is the general term for the embroidery works made mainly in Changsha of Hunan Province. Hunan embroidery is known for its exquisite needlework, bright colors and unsophisticated style. Its suede patterns are vivid with colors in proper gradation. Hunan embroidery takes traditional Chinese painting as its theme, which becomes one of its praiseworthy features. The embroiderer uses embroidery threads in over 100 colors and applies several dozen needleworks to turn out colorful and beautiful embroidery works. To make it right, he must combine painting, calligraphy and embroidery to a common end.

旗袍的纹样

纹饰是中国服饰文化中必不可少的要素，是传统符号的体现，每个都有着颇深的寓意。早期的旗袍上常见的纹样有龙、凤、祥云、蝴蝶、鱼、梅花、兰花、菊花、牡丹花等。纹样之间或绕，或穿，或并齐，组合出式样繁多的图案。

Patterns on Chi-pao

Pattern decoration is an indispensable element in Chinese clothing culture. The patterns are the symbols of traditional signs and every one of them has profound meanings. The earliest Chi-pao had patterns of loong, phoenix, auspicious cloud, butterfly, fish, plum blossom, orchid, chrysanthemum and peony, and

清末，中国开始引进西方的纺织印染机。受西方印染技术的影响，传统的锦缎、织绣等提花织物的市场逐渐缩小。印花棉布、苎麻织物、纱、人造丝等材质被大量应用于旗袍制作中。这时的旗袍纹样简单，用色单一，且多以单枝花朵或成串花朵为主，或以簇拥着的小散花为主，几何纹样、条格纹样、单色纹样等旗袍也备受青睐。

so on. The patterns wound round each other, went through each other, or aligned with each other. They combined into numerous images.

By the end of the Qing Dynasty, China had begun importing from the west looms and dyeing machines. The traditional jacquard fabrics such as brocade and embroidery were affected by Western dyeing and printing techniques and their market gradually dwindled. Instead, the

蝴蝶纹：蝴蝶围绕花朵飞舞。刺绣所用的丝线颜色对比强烈，用色大胆，颜色均匀而有晕染的效果。

Butterfly pattern: The butterflies fly around the flowers. The threads used in the embroidery have huge and bold color contrast. The colors are even and give a brush-shading effect.

镶边图案：黑底绣以折枝牡丹与袍身图案对应，颜色略深沉，用以衬托袍身的浅绿色。

Lacing pattern: The peonies with stems are embroidered on the black background. In deep colors, they correspond to the patterns on the Chi-pao to set off the light green of the Chi-pao.

牡丹纹：花朵硕大而美丽，色彩绚丽，艳而不俗。

Peony pattern: The flowers are large and beautiful. They are in gorgeous and tasteful colors.

- 清代彩绣花蝶牡丹纹旗袍（背面）

  这件旗袍是清代中晚期较为流行的款式，绿色的袍身上绣有硕大的牡丹，蝴蝶飞舞其间。

  A Chi-pao with Colorful Flower, Butterfly, and Peony Patterns Made in the Qing Dynasty (1616-1911) (Back Side)

  This Chi-pao was a popular style in the middle and late periods of the Qing Dynasty. Its main green part is embroidered with large peonies encircled by butterflies.

20世纪初,条格织物、几何图案纹织物大受欢迎,穿插着缠枝花纹、花卉图案等,显得通透而别致。20世纪三四十年代,旗袍的纹样多为含苞欲放的花朵图案,或是条纹、几何图案,颜色也由淡雅转向艳丽。多以明黄、水绿、水红等色为底,衬以硕大的花朵。大多数当代旗袍图案则较为内敛,不求富贵,只求高雅而有韵味。旗袍从远处看,展现的是一种风情;近看,才能看清上面的纹饰,于细微处见真功。

materials such as calico, ramie fabric, silk and rayon were applied to the making of Chi-pao in large quantities. At that time, the Chi-pao had simple patterns in unitary colors. In addition, most patterns were single flowers, a string of flowers, or a cluster of small flowers. The Chi-pao with geometric patterns, stripe and check patterns, and single-color patterns were warmly received.

In the early 20th century, fabrics with stripe and check patterns and geometric patterns were popular. They were often accompanied with patterns of twisting branches and flowers. The fabrics were thus made transparent and elegant. In the 1930s and 1940s, the Chi-pao often had budding flower patterns or stripe and geometric patterns. Their color also turned from quietly elegant to bold and bright. The common background colors included pure and bright yellow,

- **身穿旗袍打高尔夫球的女人**(20世纪30年代)
图中左侧女子身着橘黄底印花旗袍,色泽艳丽、明快;右侧女子着素色印花旗袍,色泽淡雅。
Two Women in Chi-pao Playing Golf (1930s)
In the picture, the woman on the left wears a printed Chi-pao with an orange background. Its colors are bright and lively. The woman on the right wears a plain printed Chi-pao and its color is quietly elegant.

light green and bright pink, decorated with giant flowers. Most modern Chi-pao patterns are less showy. They pursue elegance and good taste instead of wealth and nobility. Viewed from afar, the Chi-pao is a kind of landscape. Its patterns cannot be seen clearly until one gets close to it. The details display the true workmanship.

- 穿着几何纹旗袍的女人（20世纪30年代）

图中女子所穿旗袍带有典型的几何纹饰，图案为纵、横罗列的黑、白二色菱形块；袖口呈喇叭状。旗袍颜色素雅，具有一种简洁、轻快之感。

A Woman in the Chi-pao with Geometric Patterns (1930s)

The woman in the picture wears a Chi-pao with typical geometric patterns arranged both vertically and horizontally. They are black and white rhombus patterns. The Chi-pao has horn-shaped cuffs. In plain and elegant colors, the Chi-pao looks simple and light.

- 紫色短旗袍

这款旗袍内层为淡紫色织锦缎，淡紫色面料下部为海水纹样，外面罩有一层蓝色，增强了蓝色海水的清凉感觉；最后又被缝上透明的珠片，水滴呼之欲出。

A Purple Short Chi-pao

This Chi-pao has a light purple brocade lining, with sea wave pattern on the bottom. Outside is a blue layer, which accentuates the coolness, sent forth by the blue sea water. Finally, some transparent sequins are stitched onto the Chi-pao, further activating the already vivid water drops.

旗袍经典图案的寓意
Meanings of Typical Patterns on Chi-pao

- 清代旗袍上的"如意云头"纹

"如意云头"是清代旗袍上较为多见的装饰花纹。它一般与旗袍的滚边结合,表现为在滚边直线处向外扩展出云头纹,形成具有流动感和弧度的尾梢。一个云头纹常常始于领口条形滚边,顺延至腋下,以云头纹作为终结。云头纹体现着流畅的直线与弯曲的点的结合,极富韵律和美感。

The "Auspicious Cloud" Pattern on a Chi-pao Made in the Qing Dynasty (1616–1911)

The "Auspicious Cloud" pattern was A common decoration technique in the Chi-pao of the Qing Dynasty. Such a pattern was often used together with the edging of the Chi-pao. The cloud pattern expands outward from the straight line of the edging to form a flowing arc end. The formation of a cloud pattern often starts from a bar-shaped edging on the collarband. Then, it extends to beneath the armpit and takes shape there. The cloud pattern is the combination of smooth straight lines and curved dots, which are full of rhythm and aesthetic taste.

- "五福(蝠)捧寿团花"纹

团花正中为"寿"字,五只蝙蝠绕之而飞。此团花被用明黄线绣于紫红色绸缎之上,色彩明亮,具有祥和、长寿之寓意。五福(蝠)捧寿图案是多用于织锦缎面料的纹样。

"Five Bats Carrying Longevity Circle Flower" Pattern

In the center of the circle flower is the Chinese character for longevity. Five bats (in Chinese, the word bat is pronounced as *fu*, which means luck) fly around it. This circle flower is embroidered onto purplish red silks and satins with pure and bright yellow thread to represent harmony and longevity. The "Five Bats Carrying Longevity" pattern is often used on brocade material.

- "喜鹊报春团花"纹

团花以折枝牡丹为主体,两只喜鹊立在花枝头,一面啼鸣,一面逗弄飞舞的蝴蝶。此图案有吉祥、喜庆之寓意。

"Magpies Reporting Spring Circle Flower "Pattern

The circle flower has peonies with stems as its main contents. Two magpies (magpie is called happy bird in Chinese) stand on the flowery branch. They chirp and play with the fluttering butterflies. This pattern represents good luck and happiness.

- 园林山水纹

山水纹样是一种独特的纹样制式,带有山水纹样的旗袍较为少见。山水纹饰多以山水乡居、田园风光、亭台楼阁等为主体,写意性较强,且写实性与写意性兼具。山水纹样的寓意大多较为明晰,即欣赏山水,与山水相依。

Landscape Pattern

The Chi-pao with landscape pattern is rare. A unique pattern style, the landscape pattern mainly depicts the pastoral landscape and some beautiful manmade structures, including pavilions, terraces and towers. Strongly suggesting the freehand brushwork style in traditional Chinese painting, such pattern also contains the elements of realism painting. Most landscape patterns have a clear theme—appreciating the natural landscape and living in harmony with nature.

- **"松鹤延年"纹**

  在中国的传统文化中，松树喻示着长生，鹤则是高洁、清雅的象征。此二者合在一起喻示人如松鹤般长寿、高洁。此旗袍橘黄色织锦缎上被用黑、白二色绣出仙鹤图案，形象生动，在松林间展翅而飞。

  "Pine and Crane for Longevity" Pattern

  In traditional Chinese culture, pine symbolizes longevity and crane nobility and elegance. Combined, they bode well for a man to enjoy the same longevity and nobility as pine and crane. This Chi-pao has crane patterns embroidered with black and white threads on the orange brocade. The cranes are vividly flying among pine trees.

- **"喜上眉梢"纹**

  梅花是中国的传统名花，有高洁之寓意。此外，由于梅花呈五瓣，也象征着五福捧寿。图中喜鹊在梅梢嬉戏，谐音寓意为"喜上眉（梅）梢"。

  Magpies on Prune Tree Branches Pattern

  Symbolizing nobility, the plum blossom is a traditional famous flower in China. In addition, a plum blossom consists of five petals, which have the same meaning as "Five Bats Carrying Longevity". In the picture, magpies are playing happily on the prune tree branches, which has the same pronunciation of "*Xishang Meishao*" in Chinese.

- **"鸳鸯荷塘"纹**

  在中国，鸳鸯自古以来就被当作爱情的象征。此图案被绣于新娘嫁衣上，衬以殷红的底子，两只鸳鸯悠闲地嬉戏，回望蕴含着款款深情。

  "Mandarin Ducks in a Lotus Flower Pond" Pattern

  Mandarin duck has been the symbol of faithful love in China since ancient times. This pattern is often seen on the bridal attire with a red background. The mandarin duck couple are leisurely playing and looking at each other with affection.

- **"鸳鸯戏水"纹**

  鸳鸯经常出现在中国古代文学作品和神话传说中，一般成对出现，喻示着夫妻双宿、双栖，忠贞不渝、永不分离。

  "Mandarin Ducks Playing in Water" Pattern

  Mandarin ducks often appear in the ancient literary works, myths and legends of China, usually appear in pairs. It is a bird symbolizing the never-ending faithfulness between couples.

- **"丹凤朝阳"纹**

  凤为"百鸟之王"，是富贵、吉祥的象征。在中国古代，凤纹为贵族妇女所专用，后演变为婚庆新娘服装的主体纹样。

  "Phoenix Towards the Sun" Pattern

  Known as the King of Birds, the phoenix symbolizes wealth and good luck. In ancient China, the phoenix pattern was reserved for noble women. Later, it evolved into the major pattern for bridal attire.

• 菊花纹

菊花具有凌霜、傲雪的品质，是坚贞和高洁的象征。菊花又称"长寿花"，也是健康、长寿的象征。此外，因菊与"居"谐音，所以，它也有"安居乐业"的寓意。

Chrysanthemum Pattern

Chrysanthemum has an unyielding character. Fearing no coldness, it symbolizes faithfulness and nobility. Also called the "Flower of Longevity", it represents good health and longevity. Moreover, as chrysanthemum sounds like "live" in Chinese, it bodes well for a good and prosperous life.

• 兰花纹

兰花是中国的传统名花，以幽香著称。兰花素雅而清新、飘逸而潇洒，被看作高洁、典雅的象征。

Orchid Pattern

A traditionally famous flower in China, orchid is known for its delicate fragrance. Plain and graceful, the orchid has always been regarded as a symbol of nobility and elegance.

• 竹叶纹

竹具有志高万丈和虚心有节的品质。此外，因"竹"与"祝"谐音，竹又有庆贺之意。

Bamboo Leaf Pattern

Bamboo symbolizes a great ambition and a humble temperament. In addition, as bamboo sounds like "congratulation" in Chinese, it is also a good omen.

> 质之美

旗袍是一种非常有质感的服饰，面料十分考究，不同的质地会有不同的风格和韵味，或朴素，或雅致，或高贵，或艳丽。中国传统旗袍的面料主要为锦缎。清代末期，西方的纺织印染机和印染技术被纷纷引进国内，印花棉布、毛呢、纱等材质被大量用来制作旗袍。如今，旗袍的面料更是不拘一格，非常丰富。

> Beauty of the Texture

Chi-pao is a dress that gives a vivid impression to the viewers. It is made of top-grade materials. Different material textures give rise to different styles and charms. Some are plain or elegant while some others are valuable or gorgeous. Traditional Chinese Chi-pao was mainly made of brocade. By the end of the Qing Dynasty, dyeing machines and dyeing technologies were introduced to China. The materials such as calico, woolen cloth, and gauze were applied to the making of Chi-pao in large quantities. Nowadays, there is almost no limit to the materials that can be used to make Chi-pao.

- 清末女子正在纺纱
  A Spinning Woman at the End of the Qing Dynasty

- 中国传统的织布机
A Traditional Chinese Loom

- 迎风而站的旗袍女人（20世纪30年代）
此女子身穿棉布旗袍，下摆被风轻轻吹起，露出里面丝质的衬裙。
A Woman in Chi-pao Standing in Wind (1930s)
This woman wears a Chi-pao made of cotton cloth. The hem is slightly blown up by a gentle gust of wind. The silk underskirt can be seen.

丝绸

在中国古代，丝绸就是以蚕丝织造的纺织品。到了现代，由于纺织品原料的扩展，出现了一些人造丝制成的丝绸，于是丝绸就变成了丝织物的总称。而以纯蚕丝织造的丝绸又被特别称为"真丝"。丝绸面料大都非常柔软，丝的纤维很

Silk

In ancient China, silk cloth was a soft goods woven with natural silk. In modern times, as more textile raw materials are available, there has come into being some silk cloth made of rayon. To differentiate the two, silk made of pure natural silk is called "Real Silk". Most silk materials are very soft. The silk fiber is very thin and fine and therefore the silk clothing clings more closely to the wearer's body and looks more graceful. In addition, silk allows better heat radiation in hot weather but performs well in preserving heat in cold weather. The silk Chi-pao has a dignified and gorgeous beauty and is therefore loved by women.

There are many kinds of silk. Those often used to make Chi-pao include tapestry satin, velvet and crepe georgette.

Tapestry satin is a traditional silk fabric, which requires at least three

- **银灰色缎面旗袍**
这款旗袍被采用银灰色的缎面，营造出月夜中河塘的宁静，一朵美丽的荷花静静地开放在水中，为美丽的月光所笼罩。整件旗袍犹如一幅美丽的中国画。

A Chi-pao Made of Silvery Grey Satin
This Chi-pao is made of silvery grey satin. It is like a beautiful traditional Chinese painting. In a peaceful moonlit night, a beautiful lotus flower is quietly coming into bloom in a pond.

细，因而更服帖，更飘逸。此外，丝绸不仅具有较好的散热性能，还有很好的保暖性。以丝绸制成的旗袍给人高贵、华丽的美感，因此深受女性喜爱。

丝绸又分为很多种，常用于制作旗袍的丝绸包括织锦缎、丝绒、乔其纱等。

织锦缎是一种传统的丝织物，织造时需至少三种彩丝，花纹精致，表面光亮而细腻，质地紧密而厚实，色彩绚丽、悦目。

丝绒是割绒丝织物的统称，表面有平滑而整齐的绒毛，质地顺滑、细腻、厚重，有良好的垂感，能展示出穿着者的娴雅和宁静，适合做婚礼等盛大场合的礼服式旗袍。丝绒色彩鲜艳，表面能随光线的变化发出不同的光泽。丝绒制成的旗袍触感柔软、色泽华丽，不需要加以过多的装饰，单单是面料本身，光泽于静动间的流转变化就足以展现独特的魅力，体现出穿着者的落落大方和雍容华贵。高级丝绒面料适合被做成高领、下摆刚过膝盖的旗袍，如果在胸、领、襟处稍点缀以一些装饰，则更为光彩夺目。

kinds of color silks for its making. It has exquisite patterns, a bright, smooth and fine surface, and a compact and thick texture. It often has gorgeous colors that are pleasing to the eye.

Velvet is a general term for cut pile fabrics. On its surface are parallel and even flosses, which result in a smooth, fine, thick texture and a good drooping effect. As a result, velvet is the right material for making formal Chi-pao for grand events such as weddings. Velvet clothing can soundly demonstrate the wearer's elegance and quietness. In bright colors, velvet can reflect different luster on its surface in response to the change of the surrounding light. The velvet Chi-pao touches soft and has gorgeous colors, which allow the omission of extra decorations. With the moving of the material alone, the flows and changes of its luster are enough to reveal its unique charm. No wonder the wearers of the velvet Chi-pao look graceful and dignified. Top-grade velvet is suitable for making the Chi-pao with a high collar and a hem just covering the knees. If there are decorations on the chest, collar, and front piece, the Chi-pao will shine more brilliantly.

乔其纱俗称"雪纺",是一种质地轻薄,同时又透明的丝织物,外观清新而素雅,具有良好的透气性和悬垂性。以乔其纱制成的旗袍穿起来非常飘逸、舒适。

Also known as chiffon, crepe georgette is a thin and transparent silk fabric. With a fresh and elegant appearance, it guarantees good air permeability and a satisfactory drooping effect. Wearing the Chi-pao made of crepe georgette will be a very comfortable and graceful experience.

香云纱

香云纱是中国一种独具特色的传统面料,蕴含天然植物和矿物精华,散发着植物的清香。香云纱旗袍能展现出穿着者雍容华贵、庄重而高雅的气质,且穿得越久,旗袍越柔软、细亮,上身感觉越好。香云纱旗袍不易被虫蛀,不易腐烂,浸泡香云纱的药汁对皮肤具有消炎、消毒、平复等效用。

Gambiered Guangdong Gauze

Gambiered Guangdong gauze is a unique traditional material used in China. Containing natural plant and mineral essence, it sends forth a delicate plant aroma. The Chi-pao made of gambiered Guangdong gauze can demonstrate the wearer's dignified, solemn, and lofty temperament. The longer it is worn, the softer, finer, and brighter the Chi-pao becomes and the better the wearing experience is. The Chi-pao made of gambiered Guangdong gauze won't easily rot or be damaged by moths. In fact, the medical soup in which gambiered Guangdong gauze is soaked has many medicinal effects, including reducing skin inflammation, disinfecting and healing.

中国的丝绸文化

　　丝绸是中国传统服饰文化的象征，以精良的材质、美艳的花色和丰富的文化内涵闻名于世。中国是桑蚕丝的发源地，养蚕、缫丝技术为中国古代先民所创造，成就了中华民族文化的篇章。早在数千年前，中国就已经发明了丝绸织造及朱砂染色技术；此后，随着织机的不断改进及印染技术的不断提高，丝织品种日益丰富，并形成了一个完整的染织工艺体系。从古至今，丝绸都以柔顺的手感和亮丽的光泽备受人们喜爱。

The Silk Culture of China

Silk is the symbol of China's traditional clothing culture. It is world-renowned for its refined quality, gorgeous designs and colors, and rich cultural connotation. China is the birthplace of natural silk. The ancient Chinese created the technologies of silk worm breeding and silk reeling. They had woven a splendid chapter for the culture of the Chinese nation. As far back as several thousand years ago, the Chinese people invented silk weaving and knitting and cinnabar dyeing technique. Then, with the loom being continually improved and the dyeing technique being refined, there have been more and more silk fabrics. A complete dyeing and weaving technical system has taken shape. From ancient times to the present, silk has been loved by the people due to its smooth touch and brilliant luster.

• 正在吐丝的蚕
A Silk Worm Producing Silk

• 织锦缎
Tapestry Satin

085

Beauty of Chi-pao
旗袍之美

- 婷婷玉立的旗袍女子（20世纪30年代）

画中的女子身穿一件丝绸旗袍，柔软而贴体。

A Slim and Graceful Woman in Chi-pao (1930s)

The woman in the picture wears a silk Chi-pao, which is soft and fits perfectly.

- 喝可乐的旗袍女人（20世纪30年代）

这件旗袍以乔其纱制成，做工讲究，透出里面穿的衬裙，上身若背心状连着下身的裙子，内裙也随旗袍开衩，开衩处镶有蕾丝，透过旗袍隐约可见。

A Woman in Chi-pao Drinking Cola (1930s)

This Chi-pao is finely made of crepe georgette. The underskirt can be seen. The vest-like upper part connects with the skirt on the lower part. Following the Chi-pao, the underskirt also has slits. The decorative lace on the slits can be indistinctly seen through the Chi-pao.

西湖岸边的旗袍女人（20世纪30年代）

这位女子身穿一件丝绒旗袍，绒面细腻，花纹层次丰富，枫叶形的图案与秋季的氛围十分相符。

A Woman in Chi-pao by the West Lake (1930s)

This woman wears a velvet Chi-pao. The velvet surface is smooth and compact, with the flower patterns arranged in rich levels. The maple leaf pattern fits in well with the autumn atmosphere.

• 竹叶纹织锦缎旗袍

这款旗袍以乳白色织锦缎为基础面料，给人以在宣纸上画出的翠竹之感，在盘扣的设计上也采用了竹子造型，富有情趣，营造出了中国风格的意境美。

A Chi-pao Made of Tapestry Satin with Bamboo Leaf Pattern

This Chi-pao has milky white tapestry satin as its base material, which looks like the rice paper. Green bamboo leaves are drawn on it. The frogs are also in bamboo shape and they look vivid, presenting a Chinese-style artistic beauty.

棉布

棉布也是制作旗袍常用的面料，适合做中西结合式旗袍，能给人温柔、舒适的感觉。用于制作旗袍的棉布主要有阴丹士林布、蓝印花布等。

阴丹士林布是20世纪三四十年代盛行的一种国产棉布面料。当时，中国物资匮乏，一些高档面料的进口数量下降，价格也十分昂

- 身穿阴丹士林布旗袍的女人（20世纪30年代）
A Woman in a Chi-pao Made of Indanthrene Cloth (1930s)

Cotton Cloth

Cotton cloth is another common material for making Chi-pao, especially the one combined with Chinese and Western styles. It guarantees a gentle and comfortable feeling. The major cotton cloth varieties used for making Chi-pao include indanthrene cloth and blue cloth with a design in white.

Indanthrene cloth is a cotton cloth made in China and popular in the 1930s and 1940s. Back then, China experienced a serious lack of materials. The import of some top-grade materials dropped sharply and they became extremely expensive. To fill the gap, the cotton cloth made in China began being used to make Chi-pao. Such a trend was once popular at that time in Shanghai and the Chi-pao of this kind was loved by girl students and ordinary women in the society. Indanthrene cloth was a representative of cotton cloth made in China and it was dubbed the Patriotic Cloth. With many merits, including the pure, elegant and unfading colors, it was in fashion for a while.

The blue cloth with a design in white is a traditional folk printing and dyeing handicraft made manually in China. The blue and white porcelain patterns on it are normally made by using tie-dye and wax printing techniques and most of them are

贵，国产的棉布开始被用于旗袍制作。那时，上海一度流行以国产棉布制作的旗袍，广受女学生及社会上的平民女性喜爱。阴丹士林布是国产棉布的代表，又称"爱国布"，布料具有色彩淡雅、不易褪色等优点，风行一时。

蓝印花布是一种中国传统的民间手工印染工艺品，印花方式包括扎染、蜡染等，大多以蓝靛为染料，成品都是蓝白相间的花布，需经过手工纺织、刻版、刮浆等多道印染工艺制成。蓝印花布朴素而大方，色调清新、明快，图案淳朴、雅致，制成的旗袍非常别致。

dyed with indigo. All the finished products are blue-and-white figured cloth, which are made through several printing and dyeing procedures such as manual spinning and weaving, block cutting and paste smearing. The blue cloth with a design in white looks plain and tasteful, with fresh and vivid colors and unsophisticated and pretty patterns. The Chi-pao made of such cloth is very unique.

- 蓝印花布

A Piece of Blue Cloth with Design in White

- 现代棉布旗袍

这是一件日常穿着的旗袍，采用的是一种叫"千鸟格"的西式套装面料。用西式的面料做中式的旗袍，能增强旗袍的日常性和时尚感。

A Modern Chi-pao Made of Cotton Cloth

This is a Chi-pao for everyday wearing. It is made of a Western suit material called houndstooth. A Chinese-style Chi-pao made of Western-style materials is more suitable for everyday wearing and more fashionable.

扎染和蜡染

　　扎染和蜡染都是中国民间传统而独特的印染工艺。扎染即织物在染色过程中被部分结扎起来，使之不能全部着色的一种染色方法。这种印染工艺使布料形成深浅不均、层次丰富的色晕和皱印，既可以染带有规则纹样的普通布料，也可以染构图复杂的精美工艺品。扎染以蓝、白二色为主调，古朴而别致，宁静而淡雅，别具特色。

　　蜡染即用蜡将花纹点绘在织物上，然后放入染料缸中浸染，有蜡的地方染不上颜色，除去蜡后，就会现出美丽的花纹。蜡染造型简练，色彩单纯而明朗，装饰纹祥夸张而丰富，极具民族特色。

• 云南大理的白族妇女正在晾晒扎染制品
A Bai Woman in Dali of Yunnan Province is Airing Her Tie-dye Work

Tie-dye and Wax Printing

Tie-dye and wax printing are both traditional folk printing and dyeing techniques peculiar to China. During tie-dye, some of the fabric is tied up to prevent certain parts from being dyed. Such a technique can result in uneven dyeing and produce color halos and fold prints on different layers. The fabric can be dyed into ordinary clothes with regular patterns or exquisite handicrafts with complicated picture compositions. In tie-dye, blue and white are the major colors and they are simple, unique, quiet, elegant, and full of characteristics.

During wax printing, patterns are painted on the fabric with wax. Then, the fabric is steeped into the dye vat. Those parts covered with wax won't be dyed. Later, when the wax is removed, beautiful patterns will emerge. Wax printing produces simple patterns in pure and clear colors. There are also rich exaggerated patterns for decoration. They show rich ethnic characteristics.

• 20世纪下半叶贵州黄平苗族人的蜡染服饰
Wax Printing Works Made by the Miao People Living in Huangping of Guizhou Province, in the Second Half of the 20th Century

• 云南昭通的苗族妇女正在进行点蜡
The Miao Women in Zhaotong of Yunnan Province is Applying Wax to the Fabric

毛料

　　毛料是一种精纺毛织品，一般以纯净的绵羊毛为主，也可混以一定比例的人造纤维或天然纤维，经过多次梳理、并合、牵伸、纺纱、织造等制成。毛料具有动物毛特有的良好的弹性、柔软性和抗皱性。以毛料制成的旗袍结实、挺括、不粘身、耐脏、不容易生皱，长时间内不会变形，而且光泽含蓄、柔美，能展现出女性的知性美。毛料中有一种由混色精梳毛纱织制成的面料，质地轻薄，适合做夏季穿的旗袍。

Woollen Cloth

Woollen cloth is a worsted wool fabric. It is normally made of pure wool or wool mixed with a certain proportion of artificial or natural fibers through repeated combing, merging, stretching, spinning and weaving. Woollen cloth has as good elasticity and softness and as unique feltability and wrinkle resistance as animal hair. The Chi-pao made of woollen cloth is sturdy, stiffly upright, dirt-resisting and wrinkle-proof. It neither clings to the wearer's body nor deforms even after long-time wearing. In addition, it has a reserved and beautiful luster that can soundly demonstrate the intelligence-based beauty of women. Among all woollen cloth varieties, a material woven with combed wool yarn in mixed colors has a light and thin texture and is suitable for making the Chi-pao for summer.

- 毛料旗袍
A Chi-pao Made of Woollen Cloth

选购适合自己的旗袍
Choosing a Chi-pao Right for You

如今，旗袍已是世界流行的服饰，很多外国人，包括明星也穿起了旗袍。英国著名音乐剧表演艺术家伊莲·佩姬曾经说过："我非常喜爱中国的旗袍，它雅致而大方，很有味道，希望有一天穿着旗袍站在舞台上。"商店里的旗袍品种繁多，如何选购适合自己的旗袍是极为关键的。

Today, Chi-pao has become a popular dress in the whole world. Many foreigners, including foreign stars, wear Chi-pao. Elaine Paige, the famous British actress of theater, once said: "I am fond of Chinese Chi-pao. It is elegant and tasteful. I hope someday I can appear on the stage in Chi-pao." In shops, numerous Chi-pao varieties are available. It is important to know which one is right for you.

> 旗袍的选购

旗袍穿得好看不好看，面料很关键。因此，应尽量到一些大型的商店或专柜去购买，虽然价钱贵一些，却能保证质量和美感。

根据场合选购旗袍

在选购旗袍时，首先应注意，一定根据自己的需要来购买。比如，作为结婚礼服的旗袍，材质必须好，同时，颜色尽量鲜艳夺目，充满喜庆之感，红色是不错的选择。

作为宴会礼服的旗袍，面料应上乘，绣花丝绸、丝绒等是不错的选择。色彩和花纹不要太花哨，可带有光泽，使人看起来稳重而高雅。在颜色上可以选择白色。旗袍长度应到脚面，配一双秀气的高跟鞋，厚底鞋不适宜。

> Selection and Purchase of a Chi-pao

Whether a Chi-pao looks beautiful or not depends heavily on its material. Therefore, it is advisable that you buy your Chi-pao at some large stores or exclusive shops. Although it may be a little bit more expensive, the quality and wearing effect are guaranteed.

Choosing the Right Chi-pao for Right Occasions

You must buy the Chi-pao based on your needs. For example, the Chi-pao as a bridal attire should be made of high-quality materials and should have bright and happy colors. A red one is a good choice.

The Chi-pao as formal attire for a banquet should also be made of top-grade materials such as embroidered silk and velvet. Its color and pattern should not be too showy, but can have a luster

如果只是买平时穿的旗袍，则可随心所欲。可以选择自己喜欢的类型，也可以选择能够突出个性美的类型，但要保证穿起来舒适。在上班或外出时穿的旗袍应尽量宽松，同时线条要简洁，色彩可鲜

- 宴会旗袍礼服
  A Banquet Chi-pao

and convey the sense of solemnity and elegance. In this case, a white Chi-pao with a hem reaching your feet can be a good candidate. Don't forget to wear a pair of delicate high-heeled shoes. The pantshoes with a thick sole are not a good match.

A Chi-pao for everyday wearing can be chosen more freely. You can choose either the type you like or the one that highlights the beauty of personality. However, the principle is always to guarantee wearing comfort. If worn during work or when going out, the Chi-pao should have a loose design, simple

- 结婚旗袍礼服
  A Wedding Chi-pao

艳，可以选用棉麻、仿丝、雪纺等不易出褶皱的面料。为了行动方便，用于上班、外出穿着的旗袍的裙摆可被做成A字式，领子最好为2—3厘米的小高领、U型领、水滴领等。

lines, and bright colors. It can be made of the materials on which folds do not easily form, including cotton and linen, silkaline and chiffon. For moving around comfortably, the Chi-pao can have an A-shaped hem and a two-to-three-centimeter-high small choker, a U-shaped collar, or a water-drop-shaped collar.

旗袍的色彩

每一种颜色都有不同的特性和寓意，也都能展示出旗袍的百态风姿，体现出女性的个性和情感。旗袍的色彩运用极为精妙，色彩搭配独具匠心，除了传统的大红色，还有紫色、藕荷色、粉蓝色、橘红色等，甚至配以黑色和白色进行点缀。

红色是一种可以体现较为强烈的情感的色彩，代表热情，能使人联想到生命、太阳和火焰。中国人自古尚红，中国红是中国人在节日、婚庆中喜欢运用的主题色。同时，红色能让人感到幸福、兴奋和快乐，将它运用到旗袍上，最能表达热烈、奔放、喜悦的情感。

白色是纯洁和神圣的象征，白色服饰最能体现出穿着者高洁、清雅、沉静的品性。白色旗袍能体现出女性冰清玉洁的品性。

黄色是中国的传统色彩，在历史上曾为皇族所专用，成为尊贵、权力的象征。黄色是一种明快、活泼、亮丽的色彩，能使人联想到温暖、明亮的阳光。黄色还是一种十分引人注目的颜色，能给人带来活力，容易使人兴奋。

蓝色象征着宁静和智慧，能让人联想到天空和大海，是一种具有广阔、深邃、神秘气息的色彩。以蓝色面料制作的旗袍能很好地体现出女性清纯、理智、认真的个性。

紫色是一种充满神秘感的色彩，它由红色和蓝色混合而成，既能体现红的热情和奔放，又能体现蓝的宁静和沉稳。在中国古代，紫色被视作尊贵、奢华的象征，一般是王公贵族的服饰色彩。

绿色代表着自然、成长、希望、清新、宁静，能让人联想到大自然中的植物。绿色较为柔和，能带给人舒适、柔顺、温和的感觉。绿色旗袍最能体现女子温顺、谦和、娴静的气质。

黑色单纯而简练，是一种具有双重性的色彩，一方面象征黑暗和沉默，另一方面象征成熟、庄重和神秘。穿黑颜色的服饰，可以显示出成熟和尊贵，也可以给人神秘之感。黑色在视觉上有收缩感，穿着黑色的旗袍可以使身材显得纤长。

灰色因色调暗而沉闷，不受大多数人喜欢。但事实上，灰色格调高雅，被用于服饰，可以增强现代感。灰色旗袍富有浪漫、朦胧的美感，能很好地体现出年轻女性文静、典雅的气质，给人平易近人、大方而脱俗的感觉。

花色区别于以上几种素色，一般表现为在旗袍面料上增加各式各样的花纹和图案，款式多样，非常好看。

- **具有神秘而高贵之感的紫色旗袍**

采用紫色面料制作的旗袍给人华贵、富丽、高雅之感。但同是紫色旗袍，色调的差异会产生不同的美感：淡紫色旗袍具有浪漫、甜美、优雅、飘逸的美感；玫瑰紫的旗袍给人华丽的感觉，可展现女性的成熟、妩媚；深紫色旗袍则能够体现女性的高贵和神秘。这款旗袍以仿蕾丝棉布面料制成，融入了流行的元素，印花蕾丝棉布和真蕾丝的虚实结合增强了旗袍的生动性，非常适合日常穿着。

A Mysterious and Noble Purple Chi-pao

A Chi-pao made of purple materials looks luxurious, gorgeous and elegant. However, different hues of purple will create different senses of beauty. A light purple Chi-pao is romantic, sweet, graceful and elegant. A rose purple Chi-pao looks brilliant and can demonstrate women's maturity and charm. Still a royal purple Chi-pao can showcase women's nobility and mystique. This Chi-pao is made of lace cotton cloth, which adds to it some fashionable elements. The combination of the printed lace cotton cloth and real lace gives the Chi-pao a vivid look and makes it perfect for everyday wear.

Colors of Chi-pao

Every color has its unique traits and meanings. It can help display the diversified charm of the Chi-pao and reveal the wearer's personality and feelings. Color application to Chi-pao is a delicate work of art and the right match of colors requires an inventive mind. Besides the traditional scarlet, there are also purple, pale pinkish purple, powder blue, and tangerine, which can even be embellished with black and white.

Red is a color that can reflect intense emotions. It represents passion and implies life, the sun, and flame. The Chinese people have always adored red since ancient times. China red is the theme color during festivals and weddings. Meanwhile, red conveys to people the feeling of blessedness, excitement, and happiness. Applied to Chi-pao, red is the best color to show unrestrained warm and happy feelings.

White symbolizes chastity and holiness. White clothing can best showcase the wearer's fine characters such as nobility, elegance, and serenity. A white Chi-pao can demonstrate women's pure and clean character.

Yellow is a traditional color in China. It was reserved for the royal family in history, making it a symbol of nobility and power. Yellow is the clearest, brightest, and most vivacious color. People always associate it with the warm and bright sunlight. It is also an eye-catching color and can bring vitality and excitement to people.

Blue symbolizes serenity and wisdom. Implying the sky and the sea, it carries with it a sense of broadness, profoundness, and mystery. A Chi-pao made of blue materials can perfectly showcase women's fine characters such as purity, intelligence, sincerity, and honesty.

Purple is a color shrouded in mystery.

● **具有喜庆之感的红色旗袍**

在中国，红色象征和谐、团圆、吉祥、喜庆、福禄。正红色旗袍多用作婚礼上的新娘礼服和庆典礼服，表达幸福和喜庆；粉红色旗袍适合少女，衬托着年轻女子的温柔和可爱；紫红色旗袍适合中年女性，给人成熟、沉稳的感觉。

A Happy Red Chi-pao

In China, red means harmony, reunion, good luck, happiness, and wealth. A pure red Chi-pao often serves as formal attire for a bride or on celebrations since it conveys a sense of blessedness and happiness. A pink Chi-pao is ideal for girls. It can accentuate their tenderness and loveliness. A purplish red Chi-pao is for middle-aged women as it gives a feeling of maturity and steadiness.

It is made by mixing red and blue. Therefore, it has the passion and boldness of red and the serenity and steadiness of blue. In ancient China, purple was regarded as a symbol of nobility and luxury. It was often used by the nobles as a color for their clothing.

Green represents nature, growth, hope, freshness, and serenity. It implies the plants in Mother Nature. Green is a gentle color that conveys to people the feeling of comfort, smoothness, and mildness. A green Chi-pao can best display women's fine temperaments such as compliance, modesty and gentility, and demureness.

Pure and succinct, black is a two-sided color. On the one hand, it symbolizes darkness and silence. On the other hand, however, it represents maturity, solemnity, and mystery. Black clothing can demonstrate maturity and nobility as well as mystery and sexiness. Black shrinks the vision and therefore, a black Chi-pao can make the wearer look slender.

Grey isn't loved by most people, because people consider it is a dull color. However, the truth is that grey clothing has a graceful style and a sense of modernity. A grey Chi-pao shows a

● **具有宁静之感的蓝色旗袍**

以蓝色面料制作的旗袍能很好地展现女性清纯、理智、认真、诚实的个性。浅蓝色旗袍给人明快、淡雅、清凉的感觉，适合少女在夏季穿着，能表现出女子纯真、可爱、文静之美；天蓝色旗袍让人觉得沉静中充满生机；深蓝色旗袍适合中年女性穿着，能显现出成熟、稳重的气质；藏蓝色的亮度较低，使穿着者更显庄重、沉着。

A Quiet Blue Chi-pao

A Chi-pao made of blue materials can ideally display women's pure, wise, sincere and honest characters. A light blue Chi-pao gives a clear, delicate, cool and refreshing feeling, and is suitable for girls in summer. It brings a sense of coolness and can showcase the girls' pure, lovely, and demure beauties. An azure Chi-pao shows rich vitality in calmness. A deep-blue Chi-pao is for middle-aged women. It reflects their maturity and sedateness. A dark blue Chi-pao has low brightness and will make the wearer look more solemn and composed.

romantic and hazy sense of beauty and can perfectly showcase some fine temperaments of young ladies such as quietness and elegance. It has around it an easy-going and graceful atmosphere.

Flowery color is different from the aforesaid single colors. It is normally used on Chi-pao materials by adding all kinds of figures and patterns. It can be made into countless styles and all will be pleasing to the eye.

• 高雅的绿色旗袍

浅绿色旗袍最能体现少女的青春和活力；鲜艳的菜绿色让穿着者更显年轻、有朝气；墨绿色代表永恒、坚毅、有包容性，与任何颜色搭配，都能展现稳重、大方、温和、质朴的美感。

A Graceful Green Chi-pao

A light green Chi-pao can best showcase the youth and vitality of girls. The bright vegetable green materials can make the wearer younger and full of vitality. Blackish green represents eternity, unswerving determination, and a fine sense of accommodation. Used together with other colors, it can always display a sedate, graceful, gentle and unsophisticated sense of beauty.

根据身材选购旗袍

旗袍是国际公认的最美女装之一,与西式服装的差异就在于高领及开衩,高领使女人体态端正,开衩使女人在走动时摇曳生姿。然而,为什么今日的旗袍不如西式服装普及呢?不外乎旗袍过于贴身,活动有些不方便,同时领子过高,有约束感等。但其实,经过现代旗袍设计师的改良,这些问题已逐渐被解决。

经过改良的现代旗袍下摆呈现出多样化形式,既有西式服装的新线条,又不失旗袍传统样式的美感。

在选购旗袍时应注意,现在市场上的成衣旗袍,尺寸规格是按大众化的身材量制而成的。由于每个人身材不同,有自己独特的审美标准,在购买之前,要准确地测量出自己的"三围",即胸围、腰围、臀围。三围要与旗袍相适合或略有偏差。

很多人认为,旗袍只适合身材好的人穿。其实,只要是身材匀称的女人,穿旗袍都能体现出女性优雅的姿态,略胖或略瘦的女人都可利用旗袍的面料、装饰、花纹、色彩等来弥补

Choosing A Chi-pao According to Your Figure

The Chi-pao is internationally recognized as one of the most beautiful dress for women. Its difference from Western dresses lies in its high collar and slits. With a high collar, the woman must keep her body upright.

- **适合标准身材的女性穿着的旗袍**
  设计旗袍要根据个体差异,制作出合身、富有个性的款式。
  A Chi-pao for Ladies with Standard Figure
  When designing a Chi-pao, one should take individual differences into consideration. Thus a fit style with individual character will be made.

自己身材的不完美之处。

身材偏胖的女性宜选用竖向条纹面料的旗袍，色彩以深色为宜，质地不宜太软或太硬，在样式上可采用宽滚边形式。如果腰部较粗，滚边可延长到腰部，并且"一"字扣装饰最为适宜。尽量选择短一些的旗袍，这样会使体态显得轻盈，也给漂亮的鞋子留下了更多的展示空间。

身材过瘦的女性宜选用横纹、大花的旗袍，可选颜色浅、亮度高且较硬的面料。夏季可搭配无袖背心外套，其他季节可在旗袍外加宽大的外套、大衣等。

身材娇小的女性可以选择长襟旗袍，面料最好带有竖条纹图案，这样会在视觉上给人以修长感。旗袍开衩的大小要与身高成正比。身材越修长，开衩就要越大；身材如果相对矮小，则开衩也要小。

脖子较粗的女性可以选择无领的旗袍，这样既可以避免立领妨碍脖子的活动，又可以使脖子看起来细长一些。脖子细长的女性，可以选择立领的旗袍。

Meanwhile, the slits add swaying beauty to her when she walks. However, why is Chi-pao not as popular as Western dresses? The answer is its excessive fitness to the body that hinders movement. The high collar is also restraining. In fact, these problems have gradually been solved after the modification by modern Chi-pao designers.

Modern modified Chi-pao has a hem with diversified styles. In this way, it can have new lines as those on Western dresses and also maintain the beauty of the traditional Chi-pao design.

A Chi-pao buyer should know that the ready-made Chi-pao in the market has a size decided based on the general stature of all customers. Every woman has her own stature and aesthetic standards, and should accurately measure her own vital statistics, namely bust, waistline and hipline. These statistics should meet the size of the ready-made Chi-pao or have only minor differences.

Many people think that Chi-pao is only suited to those with a good figure. Actually, Chi-pao can better demonstrate the graceful posture of those who have a good figure. Women who are a bit of fat or thin can use the materials, decorations, patterns and colors of Chi-pao to make up for their defects.

A plump woman should choose materials neither too soft nor too hard and with vertical stripes and deep colors. Wide edging can be adopted. If she has a large waist, the edging can extend to her loin. The single-line button decoration is the best choice. In addition, shorter Chi-pao is better because it will make her look more light and there will be space for her to show off her beautiful shoes.

A thin woman should choose relatively hard materials with horizontal stripes or large flower patterns in light or highly bright colors. In summer, she can wear a sleeveless vest coat in addition; in other seasons, she can put on a loose overcoat over her Chi-pao.

A small-stature woman can choose a Chi-pao with a long front piece bearing vertical-stripe patterns. This will make her slender visually. The slits should be in direct proportion to her height. The taller she is, the larger the slits can be, and vice versa.

A woman with a thick neck can choose a collarless Chi-pao. In addition to making her neck look slim, she is free of the restriction from the collar. A woman with a slim neck can choose a Chi-pao with a standing collar.

- 身穿深色旗袍的女人
  A Woman in a Deep Color Chi-pao

脸形与旗袍选购
Face Feature and the Choice of Chi-pao

椭圆形脸 Oval Face	椭圆形脸是最适合穿着旗袍的脸形，基本上任何款式的旗袍都可以穿着。 An oval face is the best for wearing Chi-pao. It fits in well with almost any style of Chi-pao.
方形脸 Square Face	方形脸会给人以棱角分明之感，因此旗袍的领形一定圆润，这样可以弱化这种脸形的棱角。同时，尽量选择浅色旗袍，以增强女性柔美的感觉。 A square face has obvious edges. Therefore, the Chi-pao should have a round collar to ease up the rigidness brought about by it. Meanwhile, a light color Chi-pao should be chosen to enhance the mild beauty of women.
长形脸 Oblong Face	长形脸的女性在选择旗袍时，尽量选择饱满一些的领形，如高领、圆领都是不错的选择，千万不要选择长的领形。 For a woman with an oblong face, the Chi-pao with a full collar should be chosen. Both high collar and round collar are good choices. The last collar needed is an oblong one.
圆形脸 Round Face	圆形脸非常适合V形领旗袍，这样可以弱化圆形脸宽大的感觉。 A round face is suitable for the Chi-pao with a V-shaped collar, which can reduce the sense of largeness of the round face.
瓜子脸 Melon-seed-shaped Face	瓜子脸的女性也非常适合穿旗袍，不需要加以其他修饰就能将旗袍穿出美感，适合各类款式的旗袍。 A woman with a melon-seed-shaped face is also ideal for Chi-pao of any type. No further decoration is needed.
倒三角形脸 Reversed Triangle Face	上额过于宽大，下颌又过于狭小的女性尽量不要选择小领的旗袍。 If you have a wide forehead and a narrow jaw, you should not choose a small-collar Chi-pao.
菱形脸 Rhombus Face	菱形脸棱角分明，因此在穿旗袍时最好梳刘海，将上额遮盖。旗袍的领形尽量圆润。 A rhombus face also has obvious edges. It is better to wear your hair in bangs to cover part of your forehead. The collar of the Chi-pao had better be round.

根据年龄选购旗袍

　　除了身材、脸形，还可以根据不同的年龄选择旗袍。比如，年轻女性宜选色彩绚丽、花式优美的款式，以体现青春的朝气，显得活泼而俊俏。可以选择红色或黄色的旗袍，最下端最好在膝盖处。

Choosing a Chi-pao Right for Your Age

Besides figure and facial features, age is another factor that decides the choice of the Chi-pao. For a young woman, the Chi-pao in gorgeous colors and a beautiful style is better to display her youthful spirit and vivacious temperament. Red or yellow

- 2010年9月4日上午，在河南省商丘市步行街上，10多位老年妇女身着旗袍走起了"猫步"。她们中间年龄最小的56岁，最大的76岁。优雅的旗袍和精彩的表演吸引了无数年轻人拍照。（图片提供：CFP）

On the morning of September 4, 2010, over a dozen old ladies in Chi-pao staged a fashion show in the pedestrian street of Shangqiu City of Henan Province. Of them, the youngest is 56 years old and the oldest is 76. Their bright-colored Chi-pao and wonderful performance attracted countless young people to take photos.

中年女性宜选色彩较为单纯、显得富丽而高雅的款式，最好选用织锦缎面料，旗袍最好长及膝下10厘米左右之处，以体现雍容华贵之感。

Chi-pao is a good choice. Its hem had better reach the knees.

For a middle-aged woman, the Chi-pao in simple colors and a splendid style is the right choice. Tapestry satin is the best material and the hem of the Chi-pao had better reach 10 centimeters below the

- **适合老年人穿着的旗袍**

这款旗袍以黑色和绿色搭配，前襟被设计为小弧形。同质地、不同色的两种面料搭配在一起，既具有整体上的和谐美，又富有变化，给人耳目一新的感觉。

A Chi-pao Suitable for the Elderly

This Chi-pao has a black and green matchup and its front piece has a small arc design. Two materials in the same texture but different colors are used together. As a result, the Chi-pao shows the beauty of integral harmony and is full of changes, giving the viewers a brand new feeling.

- **适合中年人穿着的短旗袍**

黑色的腰带设计增加了这款旗袍的时尚元素，A形下摆增加了其日常穿着的机会。

A Short Chi-pao Suitable for Middle-aged Women

The black waistband on this Chi-pao adds to it some fashionable elements. The A-shaped hem makes it more suitable for everyday wearing.

老年女性宜选面料颜色深一些的旗袍，如墨绿色、蓝色、紫色的等。腰身尽量宽松一些，以体现庄重、文静、典雅、大方之感。

knees. Such design gives a dignified and graceful atmosphere.

For an elderly woman, the Chi-pao in relatively deep colors such as blackish green, blue and purple is better. In addition, it should have a loose design to give a solemn, demure, elegant and graceful atmosphere.

- **适合中年人穿着的长旗袍**
这是一件灰底、带有黑色图案的传统旗袍，为纯手工制作，体现了高贵而精致之美。

A Long Chi-pao Suitable for Middle-aged Women
This is a traditional Chi-pao with a grey background and black patterns. Made manually, it demonstrates the beauty of nobility and delicacy.

- **适合年轻人穿着的旗袍**
这是一款改良短旗袍，A形下摆的设计增加了其时尚感，肩部绣有牡丹花纹，半朵牡丹若隐若现，给人以无限的遐想。

A Chi-pao Suitable for Young Women
This is a modified short Chi-pao. Its A-shaped design enhances the fashionable sense. There is a peony pattern on the shoulder. The half peony is partly hidden and partly visible, leaving space for boundless reveries.

根据气质选购旗袍

气质也是选购旗袍的关键，不同的气质会演绎出不同的旗袍风情。

文静、典雅的女人比较适合传统样式的旗袍。样式简洁，最好有领、有袖；花色不宜太夸张，图案不要太跳跃；要保证面料高档、精良；对于盘扣，可采用如意形、云形等经典造型。

Choosing a Chi-pao Based on Your Temperament

Personality is another key to the choosing a Chi-pao. Different personalities will showcase different charms of Chi-pao.

A Chi-pao in traditional style is suitable for quiet and elegant women. Such Chi-pao is simple and had better have a collar and sleeves. The patterns on it should not be too exaggerated or too jumping. The materials should be the refined top-grade ones. The frog can take some classical shapes such as *ruyi* (an S-shaped ornamental object symbolizing good luck) shape and cloud shape.

The fashionable and avant-garde women can make bold innovations in choosing Chi-pao. For example, the collarless Chi-pao made of jeans is suitable for them. A row of straightforward frogs mounted by the

- 适合文静、典雅的女人穿着的旗袍

  这是一款绿色锦缎露肩旗袍，图案为手绣传统花卉，露肩、露背的设计融入了西式礼服的元素，更好地与现代时尚元素结合。

  A Chi-pao Suitable for the Quiet and Graceful Women

  This is a sleeveless Chi-pao made of green brocade. It bears traditional flower patterns embroidered manually. The shoulder and back-exposed design reveals the Western attire elements and enables the Chi-pao to better combine with modern fashion.

- **适合时尚、前卫的女人穿着的旗袍**

  这款旗袍采用的是中国传统丝绸面料，但因印染了达·芬奇的画作，具有了西式油画的美感。

  ### A Chi-pao Suitable for Fashionable and Avant-Garde Women

  This Chi-pao is made of the traditional silk material of China. However, as the works of Leonardo da Vinci are dyed on it, the Chi-pao possesses a sense of beauty peculiar to Western oil painting.

- **适合活泼、风趣的女人穿着的旗袍**

  这款旗袍采用流行的裸色，又加入了白色蕾丝和珠片刺绣，给人活泼、华丽之感。

  ### A Chi-pao Suitable for Vivacious and Witty Women

  This Chi-pao adopts the color of nude, the popular color in recent years. Added with white lace and sequin embroidery, it gives a vivacious and resplendent feeling.

　　时尚、前卫的女人选购旗袍时可以大胆革新，比如牛仔面料的无领旗袍对她们来说就非常适合，如果在腰侧镶以一排率性的盘扣，则更显个性，既有传统旗袍的元素，又富有时代气息。在面料选择上，具有金属光泽的面料或者被添加了莱卡的高弹针织面料则非常适合，一些较

waist will better display their personality. Such Chi-pao contains both traditional and modern features. For the women of this category, the materials with metallic luster or the springy knitted materials added with lycra are the right choices. Some soft or clinging materials such as gauze and silk won't fit them.

　　For vivacious and witty women,

软或贴身的面料，比如纱及绸不太适合。

活泼、风趣的女人适合穿图案较大，上面印有大花、脸谱等图案的旗袍，会演绎出一种潇洒、直率的美感。在面料的选择上，棉、麻、丝、绸都可以。旗袍的图案跳跃性要强，但面积一定不要小，比如一幅泼墨山水画，否则会显得小气。如果喜欢单色的旗袍，则盘扣必须大，还可加一些夸张的装饰，如项链、耳环等。

根据季节选购旗袍

夏季时可选购轻柔、不粘身、舒适、透气的真丝旗袍，或能吸汗的纯棉布旗袍。尤其是盛夏时节，燥热难当，应选择无领、无袖的旗袍。值得注意的是，真丝缩水率较高，因此在

the Chi-pao with large flower patterns or facial images is the right choice since it gives them an unconventional and candid sense of beauty. As for the materials, cotton cloth, linen and silk are the right ones. The patterns on the Chi-pao should be jumping and big, for example, as big as an entire splashed-ink landscape. Otherwise, it won't be tasteful enough. If a single-color Chi-pao is chosen, the frogs on it should be large ones. Some exaggerated decorative necklaces, earrings and other jeweleries can also be adopted.

- **适合夏季穿着的旗袍**
  这是一款银色的梅花织锦缎旗袍，胸前透明的丝起到了良好的装饰作用，梅花形的小盘扣和面料的梅花图案上下呼应，整体典雅而大方。

  A Chi-pao Suitable for Summer
  This is a silvery Chi-pao made of tapestry satin with plum blossom patterns. The transparent silk on the chest plays a good decorative role. The small plum-blossom-shaped frogs and the plum blossom patterns on the material work in concert with each other and make the Chi-pao an elegant and tasteful dress.

购买真丝旗袍时，应以选比实际需求大一码的为宜。

春、秋季旗袍一般采用化纤或混纺织品面料，如优质丝绸或高级丝绒。这些织品虽然吸湿性和透气性都较差，却比棉织品挺括，在春、秋两季穿最为合适。

Choosing Different Chi-pao for Different Seasons

The Chi-pao for summer is normally made of real silk, which is light, comfortable, and good for ventilation, and won't cling to the body. A Chi-pao made of pure cotton cloth is another good option. As for the design, the Chi-pao for summer had better be collarless and sleeveless. It is noteworthy that real silk has a high shrinkage. Therefore, if you want to buy a real silk Chi-pao, you should choose the one one size larger than what you desire.

The Chi-pao for spring and autumn is normally made of chemical fibers or blend fabrics such as high-quality silk or top-grade velvet. Although these fabrics have poor hygroscopic properties and breathability, they remain upright longer than cotton fabrics and are the ideal choice for spring and autumn.

- 适合春、秋两季穿着的旗袍

这款旗袍采用红、绿色双鱼传统面料，喻示"年年有余"，具有中国传统的美好寓意；精致的盘扣更是点睛之笔。

A Chi-pao Suitable for Spring and Autumn

This Chi-pao is made of the traditional red and green double-fish material, which in Chinese means "Enjoying Abundance Every Year". The exquisite frogs make the Chi-pao even more beautiful.

鉴定真丝旗袍面料的方法
Methods to Identify Real Silk Chi-pao Material

1.品号识别法——中国产绸缎实行由5位阿拉伯数字组成的统一品号，购买时可以看品号的第一位，它代表织物的材质号：全真丝织物（如桑蚕丝、绢丝）为"1"，化纤织物为"2"，混纺织物为"3"，柞蚕丝织物为"4"，人造丝织物为"5"。

1. Identification by product number—Every kind of silk fabric made in China has a unified product number composed of five Arabic numerals. The first numeral tells about its material: 1 for All Silk Fabrics (such as mulberry silk and spun silk), 2 for Chemical Fiber Fabrics, 3 for Blend Fabrics, 4 for Tussah Fabrics and 5 for Rayon Fabrics.

2.价格识别法——真丝制成的旗袍价格会比较高，大约是化纤、仿真丝绸缎的两倍。

2. Identification by price—The Chi-pao made of real silk fabrics will be more expensive. Its price will be twice that of the Chi-pao made of chemical fiber fabrics and silkaline fabrics.

3.光泽识别法——真丝有吸光的性能，光泽幽雅、柔和，有珍珠般的光亮；仿真丝织物手感较柔软，但表面发暗，无珍珠光泽；化纤织物光泽明亮、刺眼。

3. Identification by luster—Real silk absorbs light and shows quiet and gentle luster, like that of a pearl. Silkaline fabrics touch soft, but have a dim surface and lack the luster of a pearl. Chemical fiber fabrics have bright and dazzling luster.

4.手感识别法——真丝手感柔和、飘逸，丝线较密，用手抓会有皱纹，纯度越高、密度越大的丝绸手感也越好；而化纤织物手感较硬。

4. Identification by touch—Real silk touches gentle and smooth. Its thread is relatively dense and will crease once scratched by hand. The silk with higher purity and greater density will touch better. In comparison, the chemical fiber fabrics touch harder.

5.燃烧法——抽出部分纱线进行燃烧。真丝在燃烧时看不见明火，会散发出一种烧毛发的气味，丝呈黑色微粒状，能用手捏碎；仿真丝燃烧时会起火苗，同时会散发出塑料味，火熄灭后边缘会留下硬质的胶块。

5. Identification by burning—Extract some yarn and burn it. Real silk won't burn into open fire, but will send forth a burnt hair smell. The ashes are black particles that can be crushed with fingers. The silkaline fabric will burn into open fire and send forth a plastic smell. After burning, some hard rubber-like remains can be found on the edge.

选购旗袍的几大要素
Major Elements for Attention in Choosing a Chi-pao

裁剪 Tailoring	裁剪对于一件旗袍非常重要，要注重合体。如果一件旗袍不合体，那么就会暴露穿着者身材的不完美之外，体现不出女性的曲线美。 Tailoring is essential for making a Chi-pao fit. If it does not fit, the defects of the wearer's figure will be exposed and her feminine line of beauty won't be fully displayed.
手工刺绣 Handmade Embroidery	一件好旗袍的价值往往在于手工刺绣的精致。制作精美的刺绣时，会根据旗袍的样式、面料和花纹来设计图案并选线。 The value of a good Chi-pao often lies in the delicacy of the handmade embroidery. Good embroidery has the pattern and thread chosen based on the style, material and pattern of the Chi-pao.
面料 Material	面料的柔软度会直接影响旗袍与身体的贴合度。一般来说，传统的绸缎会较好地掩饰身材的不完美；真丝面料则较软，穿起来更舒适。现在有很多女性喜欢莱卡等新式面料，这也是不错的选择。 Softness of the material will directly decide how well the Chi-pao will fit the body. Normally, traditional silk can soundly cover up some defects in one's figure. Real silk material is relatively soft and wears more comfortable. Now, many women like new materials such as lycra, which is indeed good choice.
花纹 Pattern	旗袍的花纹最好不要太花哨，而且要根据场合来选择。 A Chi-pao should not have too showy patterns. The patterns should be chosen according to the occasion.
领形 Collar Shape	旗袍的领形很多，最好根据自己的脸形来选择。 There are many collar shapes available for Chi-pao. A wearer had better choose one according to her own face features.
袖形 Sleeve Shape	旗袍的袖形，最好根据自己的臂长来选择。 The shape of the sleeves of the Chi-pao had better be chosen according to the wearer's arm length.

开衩 Slit	开衩可以展现女人的妩媚，但腿形不完美的女性最好不要选购开衩过大的旗袍。 Slits can display a woman's beauty of charming. However, large slits are not suitable for those with imperfect legs.
开襟 Front piece	可根据自己的喜好及胸部的丰满与否来选择旗袍的开襟。 The front piece of a Chi-pao can be chosen according to the wearer's preference and the size of her bust.
裙摆 Hem	旗袍裙摆的长度从膝下到膝盖往上几厘米处不等，一般依据年龄被划分。现在一般的礼服旗袍多采用前短后长、下摆呈弧线状的优美的鱼尾形裙摆。 The length of the hem of a Chi-pao ranges from below the knees to several centimeters above the knees. Women in different ages can choose different hem lengths. Current formal Chi-pao normally has a beautiful fish-tail-shaped hem. Such curved hem is shorter in the front and longer in the rear.
盘扣 Frog	判定一件旗袍做工的好坏，可以从盘扣入手。比如盘扣与领形、花纹是否搭配，颜色是否和谐，做工是否精良等。 Judgment of the quality of a Chi-pao can begin with frog. For example, whether or not the frog matches well with the collar shape and pattern; whether or not its color fits; whether or not it is made excellently.

> 旗袍的定制

旗袍制作的每一个环节都很重要，从前期的准备到后期的定型，每一分、每一寸都关系着旗袍的最

> Tailor-made Chi-pao

During the making of a Chi-pao, every step is very important. All details of the process, from early-stage preparation to later-stage setting, will decide the final shaping of the Chi-pao. If conditions allow, one should have her own Chi-pao tailor-made. Only the tailor-made Chi-pao will fully fit her figure and best demonstrate her personality and charm.

- 定制的手工旗袍

  这是一款玫红色的大花织锦缎传统旗袍，料子显得华贵而大方，大开襟，纯手工制作的盘扣体现了这件旗袍制作的精良。

  A Tailor-Made Chi-pao Made by Hand

  This is a traditional Chi-pao made of rose large-flower tapestry satin, which is luxurious and tasteful. The Chi-pao has a large front piece and handmade frogs. All these demonstrate its excellent making.

终造型。如果条件允许，要定制适合自己的旗袍。因为定制的旗袍才能完全适合自己的身材，更好地展现自己的个性和妩媚。

定制旗袍与成衣旗袍的差别

1.版型不同。成衣旗袍版型规矩，无法兼顾各种体型，因此不可

Differences Between the Tailor-made Chi-pao and Ready-made Chi-pao

1. Different model formats. The ready-made Chi-pao has a fixed model format and cannot fit all body configurations. It therefore cannot fit perfectly. However, a tailor-made Chi-pao is made specifically according to the wearer's stature and decorate her figure. As a result, the tailor-made Chi-pao looks much more perfect.

2. A waist-strengthening effect. A tailor-made Chi-pao, especially a top-grade one, can straighten up the waist and back of the wearer. In addition, its hem won't fold outward.

3. Diversified collar and sleeves. For a tailor-made Chi-pao, its collar and sleeves won't take shape until after over a dozen special procedures. Therefore, the collar and sleeves are stiff, in good

- **红色碎花纹旗袍**
一件旗袍的诞生凝聚了制作者的心血，经过了周密的量体、巧妙的设计、精心的剪裁和缝制，以及镶、滚、嵌、盘、绣等装饰工艺。当一件旗袍最终呈现在人们眼前时，细细观之，每一个细节都透着精understand，都有严格的要求。

A Red Chi-pao with Small Flower Patterns

A Chi-pao is the fruit of its maker's hard work such as precise measurement, ingenious design, painstaking tailoring and sewing, and decorative processing such as lacing, edging, embedding, coiling and embroidering. Viewed carefully, every detail is refined and has been processed in accordance with strict requirements.

能做到十分贴身。定制旗袍时，制作者会为每一位顾客量身打版，以修饰她们不同的身材，使她们在穿旗袍时更显完美。

2.手工定制的旗袍有拔腰的效果。高档的手工旗袍能使穿着者腰身挺拔，且下摆不会外翻。

3.有着多样的领子和袖子。手工制作旗袍时，领子和袖子都要经过十几道特殊工艺才能完成，所以做出来的领子和袖子非常挺括、有形，曲线也十分自然。

4.暗扣牢固。成衣旗袍的暗扣大多是按扣，不太牢固。定制的旗袍则采用按扣与钩扣相结合的方式，更加牢固。

5.面料特殊。一些特殊的面料只有经过手工制作，才能达到最佳效果，比如绒、沙绒、蕾丝等。

6.面料上带有绣花图案。手工刺绣远比机制的成衣旗袍上的图案精细，颜色更丰富，画面更生动，更能彰显穿着者的个性和品位。

定制旗袍的步骤

以传统手工制作一件旗袍，需要进行量体、定款式、画纸版、裁

shapes, and with natural curves.

4. Firm covert buttons. The covert buttons of a ready-made Chi-pao are mostly snap-fasteners, which are not firm enough. A tailor-made Chi-pao combines a snap-fastener with a hook-fastener and wears more securely.

5. Special material. Some special materials can achieve the best effect only by hand tailoring. These include velvet, sand pile and lace.

6. Embroidered patterns on the fabric. Handmade embroidery on tailor-made Chi-pao is far more refined than the machine-made ones on the ready-made Chi-pao. With richer color and more vivid patterns, it can better showcase the wearer's personality and taste.

Steps for Making a Tailor-made Chi-pao

The traditional way of making a Chi-pao by hand contains many steps, including body measurement, style determination, drawing a paper mold, cutting out the paper mold, material calculation, drawing on the material, cutting out the material, blocking of specific parts, sewing dart and panel, sewing shoulder seam and side seam of the cloth piece, making collar,

纸版、算料、画面料、裁面料、归拔特定部位、缝合省道、嵌条、缝合衣片的肩缝和侧缝、做领子、上领子、合袖片、上袖子、各部位滚边（开衩处锁线）、制作底边、做盘扣、钉盘扣、钉暗扣等步骤。其中，仅量体就需要测量全身36处的尺寸。

mounting collar, sewing sleeve pieces together, mounting the sleeves, edging of different parts (sewing the side of the slit), making the bottom side, making frog, sewing on frog, and sewing on covert buttons. Of these steps, body measurement shall be done at 36 body parts.

To have her Chi-pao tailor-made, a woman just needs to do two things: taking her own body measurements and choosing a style. Then, she can leave the rest work to the tailor. She can choose the style and decide on the collar shape, sleeve length, front piece, and height of the slits based on the texture of the material and her personal preference.

Every woman's figure is unique. Chi-pao is a dress clinging to the body. Its size specifications are the important indices in choosing a ready-made Chi-pao or having one tailor-made. Therefore, a woman must precisely measure her body parts before purchasing a Chi-pao or having her own Chi-pao tailor-made.

- **定制的旗袍**
定制的旗袍更加合体，更能凸显女性的优美曲线。
Tailor-made Chi-pao
A tailor-made Chi-pao fits better and can better show the attractive curves of a female.

定制旗袍时，只需让专业的裁缝量出自己身体各部位的尺寸，再选定款式即可，其余工作可由裁缝完成。量好身形，即可根据旗袍面料的质地和个人爱好定款式，以及领形、袖长、衣襟、开衩高低等旗袍各部位的细节。

每个女人的身材都有特殊性，而旗袍又是紧身的服装，尺寸规格是选购和定制旗袍的重要指标。所以，购买或定制旗袍前必须被准确地测量出自己的身体尺寸。现代手工定制旗袍的过程则相对简化，但一般也需要量出至少18个重要部位的尺寸，才可将旗袍做得完全贴身。

1. 身高（代表旗袍的"号"）：从头顶部垂直到脚后跟的距离。

2. 旗袍长：从颈侧点经乳峰点（胸部最高点）到旗袍末端的距离。

3. 套装旗袍衣长：从颈侧点经乳峰点往下量至所需长度。

4. 套装旗袍裙长：从腰围线经臀围线往下量至所需长度。

5. 旗袍领高：从颈侧点经脖颈往上量至所需长度（一般为3 cm—5 cm）。

6. 旗袍领围：经过第七颈椎点和

The modern way of making a Chi-pao is relatively simplified. However, at least 18 vital parts should be measured in order to make a Chi-pao fully fit.

1. Height (represented by the "number" of the Chi-pao): the distance from the top of the head to the heel.

2. Length of Chi-pao: the distance from the neck side to the end of the Chi-pao via the highest point on the breast.

3. Length of Chi-pao suit: the length measured from the neck side downwards to the desired position via the highest point on the breast.

4. Length of Chi-pao skirt: the length measured from the waistline downwards to the desired position via the hipline.

5. Height of the collar of the Chi-pao: the length measured from the neck side upwards to the desired position via neck (normally between 3 centimeters and 5 centimeters).

6. Collar girth of the Chi-pao: the girth via the seventh cervical vertebrate point and neck side point.

7. Chest breadth of the Chi-pao: the distance between the two front armpit points.

8. Knee length of the Chi-pao: from waistline to the knee.

9. Length of the back of the Chi-pao: the length from the hind neck point to the

颈侧点一周的围度。

　　7.旗袍胸宽：两前腋点之间的距离。

　　8.旗袍膝盖长：腰围线到膝盖的距离。

　　9.旗袍后背长：后颈点至后腰节点的距离。

　　10.旗袍胸高：保持立姿时颈侧点至乳峰点的距离。

　　11.旗袍胸围：在衬衫外沿腋下，通过胸部最丰满处平衡围量一周。这是紧胸围尺寸，还应按品种要求加放所需松度。

　　12.旗袍总肩宽：左肩外端经后脖根至右肩外端的距离，可根据所选择的款式增加或减少肩宽尺寸。

　　13.旗袍袖长：左肩骨外端至手腕的距离，可根据所选择的款式增加或减少袖长。

　　14.旗袍袖口围：短袖款式，量手臂围一周；长袖款式，量手腕围一周。可根据所选择的款式增加或减少袖围。

　　15.旗袍腰节：由前身左侧脖根处（肩领点）经过胸部最高处量至腰间最细处。

　　16.旗袍腰长：从腰围线到臀围线的距离。

　　10. Chest height of the Chi-pao: the distance from the neck side to the highest point on the breast in a standing position.

　　11. Bust of the Chi-pao: measure the fullest circle around the chest via the armpit. This is a tight bust size. The actual size can be larger based on the material variety.

　　12. Total shoulder width of the Chi-pao: measured from the outer point of the left shoulder to the outer point of the right shoulder via the back neck, increase or decrease shoulder width based on the chosen style.

　　13. Sleeve length of the Chi-pao: measured from the outer end of the left shoulder to the wrist, increase or decrease sleeve length based on the chosen style.

　　14. Cuff girth of the Chi-pao: for short-sleeve style, measure one circle of the arm; for long-sleeve style, measure one circle of the wrist, increase or decrease cuff girth based on the chosen style.

　　15. Waist of the Chi-pao: measured from the left-side nape (shoulder collar point) to the thinnest part of the waist via the highest point on the breast.

　　16. Waist length of the Chi-pao: the distance from waistline to hipline.

　　17. Waistline of the Chi-pao:

17.旗袍腰围：在腰间最细处围量一周，并按款式要求放出松度。

18.旗袍臀围：沿臀部最丰满处水平围量一周，并按要求放出松度。

measure one circle around the thinnest part of the waist, and give looseness based on the style requirements.

18. Hipline of the Chi-pao: measure one circle around the fullest part of the hip, and give looseness based on the requirements.

定制旗袍尺寸测量小贴士

1.测量时最好穿紧身内衣。

2.量身之前，需要在腰围最细处和臀围最宽处各系一条前后平行的绳子，并以绳子为标准线量取数据。量体的顺序一般是先量长，再量宽，最后确定体型轮廓。

3.测量时保持身体自然放松，端正姿态。

4.测量时一定保持皮尺水平、松紧适当，不能过松，也不能过紧。

5.如自己无法测量准确，可记下以上需测量的部位，去成衣店请专业裁缝帮忙测量。

Tips on Size Measurement for the Making of a Tailor-made Chi-pao

1. Wear close-fitting underwear during the measurement.

2. Before the measurement, tie a rope around the waist on the thinnest part and another rope around the hips on the widest part. The two ropes are parallel to each other. They serve as standard lines for taking the measurement data. Length is measured before width. Finally, the stature contour is determined.

3. During the measurement, one should keep a natural, relaxed, and correct posture.

4. During the measurement, one should keep the tape even and in proper tightness.

5. If one cannot take the measurement herself, she can write down the aforesaid must-measure parts and seek help from a professional tailor.

> 旗袍的配饰

　　配饰对旗袍来说也是十分重要的。旗袍的搭配形式多种多样，戴首饰、提手袋、拿扇子、撑阳伞，

> Adornments for Chi-pao

Adornments are very important for a Chi-pao. There are diversified ways to adorn a Chi-pao, including wearing ornaments, carrying a handbag, holding a fan or carrying a parasol. A pair of beautiful shoes and lovely silk stockings is also

- 油画《持伞的旗袍女子》（作者：姜迎久）
阳伞是旗袍女子常用的遮阳工具，同时具有装饰的作用。中国传统的阳伞一般由油纸、油布或细绢制成，以竹为骨架，伞面上多绘印着各种花鸟、山水等精美图案。

The Canvas *a Woman in Chi-pao Holding a Parasol* (by Jiang Yingjiu)

A parasol is often used by a woman wearing a Chi-pao. It plays a decorative role. A traditional Chinese parasol is normally made of oilpaper, oilcloth or fine silk, with a piece of bamboo as its skeleton. On the parasol are beautiful patterns such as flowers, birds and landscapes.

再选一双漂亮的鞋子，配上一双妩媚的丝袜……百变的搭配为旗袍女子增添了别样的风彩。

advisable. The ever-changing adornments will add unique charm to the wearer of a Chi-pao.

手套：手套能让穿旗袍的女人显得非常典雅。适合穿旗袍佩戴的手套多由棉布、绢纱或涤纶制成，透气性较强，表面多有刺绣，装饰性很强。与旗袍相配的多为长手套，手腕处往往有很多修饰性成分：有的点缀以花边；有的开出一条细衩让细嫩的手腕若隐若现；有的缀以明线，或饰以波浪边。

Gloves: A pair of gloves can make the woman wearing a Chi-pao look elegant. The gloves suitable for a Chi-pao are usually made of cotton cloth, silk gauze or polyester fiber. Bearing embroidery, they have good breathability and great decorative properties. Normally, long gloves with decorations on the wrist are suitable for Chi-pao. Some gloves have laces and some others have thin slits to vaguely expose the wearer's thin and tender wrists. Still some have clear lines decorated with wave-shaped lace.

手袋：手袋是旗袍女子出门时必不可少的装备。

Handbag: A handbag is an indispensable adornment for a lady wearing a Chi-pao.

- **身穿蓝色印花旗袍的女人（20世纪40年代）**

  这件长旗袍给人清爽、端庄的美感。旗袍长及脚踝，腰身紧而窄，两边开衩，显示出女性修长的身材。旗袍的外形呈流线形，线条简洁、明朗。

  A Woman in a Chi-pao Made of Blue Printed Cloth (1940s)

  This long Chi-pao gives a fresh and dignified sense of beauty. Its hem reaches the ankles and it tightens up at the waist. With slits at both sides, the Chi-pao finely demonstrates the woman's slim figure. The Chi-pao has a streamlined contour with simple and clear lines.

妆容

　　化妆对于女人是必不可少的，尤其在穿旗袍时。早在20世纪20年代，各种各样的香粉就已经出现了。那时，女人在上妆之前，会先在脸上敷上一层雪花膏或珍珠霜，然后抹上鹅蛋粉，再涂一层扑粉，接着描眉、涂口红、擦胭脂，以最漂亮的面容映衬妩媚的旗袍。

- **身穿中袖旗袍的女人（20世纪30年代）**
这位女子头戴礼帽，胸前系着丝巾，脸上妆容艳丽，仪态万千。

A Woman in a Medium-sleeve Chi-pao (1930s)

This woman wears a hat and has a silk scarf on her chest. With gorgeous makeup on her face, she looks incomparably graceful.

Makeup

Makeup is indispensable for a woman, especially the one wearing a Chi-pao. As far back as the 1920s, various face powders had appeared. At that time, women would apply a layer of vanishing cream or pearl cream on their face before using cosmetics. Then, they would apply some powders before tracing eyebrows, painting lipstick, and applying rouge. All women want to face others with their most beautiful look, to suit the graceful Chi-pao.

　　Nowadays, every woman will make-up before going out. However, the makeup should match the style of the Chi-pao so as to display the unique charm of the Chi-pao.

　　When wearing a Chi-pao, it is unwise to put on gay and heavy makeup. Except for some exaggeration on eyelashes and eyebrows, the makeup in other parts should be natural, clean and vivid. The color of the lipstick should work in concert with the color of the Chi-pao. The Chi-pao wearer can paint her eyes into a pair of tilted eyes by painting an elongated and upturning line at the corner of the eye. The lips should be as full as possible and the cheek rouge had better be applied upwards obliquely to

当今社会，几乎每个女人出门前都会化妆，但穿旗袍时的妆容要与它搭配，这样才能呈现出旗袍独 give the face a three-dimensional appeal. The eyebrow can be painted into a curved willow leaf shape or just made up based on its natural shape.

The hairdo had better be a Chinese-style classical bun to show the beauty of feminine elegance. The hair can be worn in natural or ancient-style

- **留中长波浪发型的旗袍女人（20世纪30年代）**
  20世纪30年代，西方烫发技术传入中国，烫发普及开来，样式以波浪式为主。小卷烫发是那时最为流行的发型，卷发贴着脸颊而下，极为柔美。当时的标准美人的主要特征是樱桃小口、柳叶眉、小卷烫发夹于耳后，并露出别致的耳环。此画中坐在石墩上的女子留着波浪发型，长度及肩，娇美动人。

A Woman in Chi-pao with a Long-wave Hairdo (1930s)
In the 1930s, the Western perm technique was introduced into China and soon, perm was spread all over China. The wave-shaped perm was the main style at that time. The most popular hairdo was the small-roll perm. The wavy hair dangled down along the cheek, giving the woman a gentle beautiful look. At that time, the standard beauty is the one with a small cherry-like mouth, willow-leaf eyebrow, small-roll perm and a pair of unique earrings. In this picture, the woman sitting on the stone stool wears a wavy hairdo, which reaches her shoulder and makes her beautiful and graceful.

有的韵味。

　　穿旗袍时切忌浓妆艳抹，除了可以对睫毛和眉毛稍做修饰，其他部位的化妆都应追求自然、干净而有质感。唇彩的颜色要与旗袍的颜色呼应。眼妆可以采用丹凤眼的画法，将眼尾拉长，使之呈上翘状。尽量将唇形画得丰满。擦腮红时最好采用斜向上的扫法，以提升面部的立体感。眉妆可采用弯弯的柳叶眉，也可采用保持天然眉形的妆容。

　　发型最好采用中式的古典盘发和发髻，以展现出女性的古典美和优雅。可以不留刘海，也可采用自然的偏分或复古的波涛式刘海，这都是比较适合穿旗袍的发型。头饰要以小型饰品为主，尤其是珍珠类的发饰，这样更能突出旗袍整体造型的古典美。

wave-shaped bangs or without bangs as both are suitable for Chi-pao dressing. Small headwear is desired, especially the pearl-type hair decorations, which can better accentuate the classical beauty of the Chi-pao.

- **现代旗袍女子的妆容**
  现代旗袍女子的妆容不宜浓艳，要恰到好处，尤其是唇彩和眼影的颜色，要与旗袍颜色协调。

  Makeup of a Woman in a Modern Chi-pao
  A woman wearing a modern Chi-pao should not bear heavy makeup, but an appropriate one. The color of her lipstick and eye shadow must be in harmony with the color of her Chi-pao.

首饰

首饰是旗袍的"闺中密友",想把旗袍穿出韵味,首饰必不可少。一般来说,女人的首饰主要有耳环、项链、胸花、手镯、戒指等。首饰的材质有玉、翡翠、玛瑙、珍珠、景泰蓝(一种瓷、铜结合的独特的工艺品,多为蓝色,故名。造型独特,制作精美,图案庄重,光彩夺目,具有鲜明的中国传统特色)等,都非常适合与旗袍搭配。

Ornaments

Ornaments are intimate friends to Chi-pao. To make her Chi-pao graceful, a woman must wear ornaments. Generally speaking, women's ornaments include earrings, necklaces, broochs, bracelets, rings, and so on. The ornaments are made of jade, emerald, agate, pearl and cloisonné. (Cloisonné: a unique handicraft made of porcelain and copper and mostly in blue. With special modeling, exquisite making, solemn pattern and splendid colors, it shows distinct traditional characteristics of China.) They are all good matches for Chi-pao.

- **戴珍珠项链的现代旗袍女子**

  珍珠项链是穿旗袍时绝妙的搭配。因为珍珠的光泽比较含蓄,与旗袍的风格非常协调。从形状和颜色方面来讲,珍珠是纯洁、真诚、圆满的象征,尤其是那洁白似月的银光,给人朦胧、神秘之感。

  A Woman in a Modern Chi-pao Wearing a Pearl Necklace

  A pearl necklace matches a Chi-pao perfectly. The luster of pearl is rather reserved and gets along well with the style of the Chi-pao. The shape and color of a pearl symbolize purity, sincerity and perfection. In particular, its whiteness is very much like the silvery light of the moon, giving the viewers a vague and mysterious sense.

发型：穿旗袍适合盘发，但也可以不盘，保持长发自然、垂顺的状态或烫发都可以。但无论梳什么样的发型，都最好戴一件发饰，发饰的颜色应是旗袍颜色的同类色，并且不要太抢眼。

Hairdo: Bun is the best hairdo for Chi-pao. However, other hairdos such as natural long hair or perm are also acceptable so long as a hair decoration is used. Such hair decoration should not be obtrusive and can be in the same color as that of the Chi-pao.

耳环：用于搭配旗袍的耳环愈小愈佳，或珍珠一粒，或金银一点，若隐若现即可。

Earrings: To match a Chi-pao, the smaller earrings are a better choice. A single pearl or a small bit of gold and silver will do.

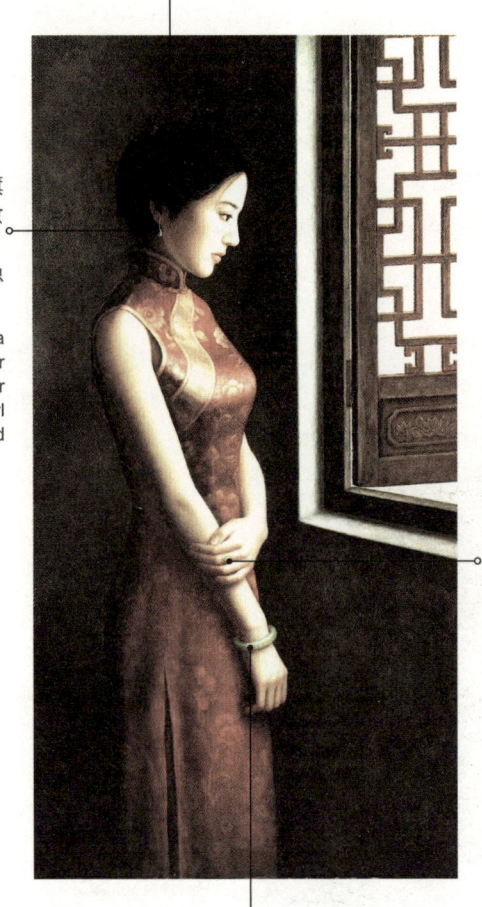

戒指：戒指是穿着旗袍时很好的点缀之物，一般由翠玉、紫晶、水晶、玛瑙、金银镶嵌珠宝等制成。

Ring: A ring is also a good decoration for the Chi-pao. Normally, a ring made of jade, amethyst, crystal, agate, and gold or silver embedded with jewels is a good choice.

手镯：手镯是搭配旗袍的一个重要物件，多以金、银、玉石、珍珠等材质制成，风格多样。穿旗袍时比较适合戴玉手镯和翡翠手镯。

Bracelet: Another important article to match a Chi-pao, the bracelet is usually made of gold, silver, jade and pearl in diversified designs. In comparison, the bracelet decorated with jade and emerald is more suitable.

耳饰：旗袍配耳饰是非常适合的，精致的耳饰闪烁于发际，美艳无比。适合穿旗袍时佩戴的耳饰有珍珠、钻石、红宝石、金、银等材质。

Earrings: Earrings are good ornaments for a Chi-pao. The exquisite earrings twinkle on the ears and are extremely beautiful. The earrings suitable for the Chi-pao are those made of pearl, diamond, ruby, gold and silver.

披肩：穿旗袍时，围巾也是不错的搭配。围巾一般分披肩、围脖、包头巾等，搭配旗袍的围巾还包括色彩缤纷的小型饰巾（如方巾、长条巾等）。

Shawl: A scarf is another nice decoration for a Chi-pao. The scarf normally includes a shawl, neckerchief, and turban. The colorful small-sized handkerchiefs (such as the square scarf and long scarf) are the right decorations for a Chi-pao.

- **时髦的旗袍女子（20世纪40年代）**
  图中女子身穿的白色印花旗袍图案简洁、雅致而明快，两边开衩，线条流畅。一件西式披肩被罩在外面，既保暖，又美观。

  A Fashionable Woman in Chi-pao (1940s)
  This woman wears a white printed Chi-pao with simple and elegant patterns. It has slits on both sides. The lines are smooth and simple. A Western shawl is put on top of the Chi-pao. It plays double roles: maintaining warmth and adding beauty.

中国传统首饰

中国传统的首饰融合了几千年来世代手工匠的智慧结晶，种类繁多，样式精美，是现代珠宝设计的艺术基石。中国传统首饰主要包括耳环、项链、胸花、手镯、戒指等，材质有玉、翡翠、珍珠、金、银、铜等。这些首饰历史悠久，早在原始社会，中国女性就已懂得用耳环等饰物装饰自己了。

Traditional Ornaments of China

Traditional ornaments of China are the fruits of the wisdom of craftsmen during the past several millenniums. In diversified varieties and exquisite styles, they are the artistic foundation for modern jewelry design. Traditional Chinese ornaments mainly include earrings, necklaces, broochs, bracelets and rings. They were normally made of jade, emerald, pearl, gold, silver and copper. They had a very long history. In fact, as far back as primitive society, women in China knew how to decorate themselves with earrings.

- 镶金镯（清）
镶金的饰品光彩夺目，经久不褪色，因而银珍贵。
A Bracelet Mounted with Gold (Qing Dynasty, 1616-1911)
The gold-inlaid ornaments are bright-colored and dazzling, so they are very precious.

- 金镶玉手镯（清）
金镶玉的材质使手镯既有玉的温润，也有金的富贵。
A Gold Bracelet Mounted with Jade (Qing Dynasty, 1616-1911)
The gold bracelet mounted with jade has both the mildness of jade and the wealthy appearance of gold.

- **镂空葫芦形金耳环（清）**

此款镂空耳环玲珑剔透，做工精细，是清代贵族妇女的饰品。

A Pair of Hollowed out Gold Earrings in Gourd Shape (Qing Dynasty, 1616-1911)

This pair of hollowed-out earrings is dainty and exquisite with refined workmanship. It was once belonging to a noble woman in the Qing Dynasty (1616-1911).

- **翡翠镶钻项链**

此项链的珠子翠绿、晶莹剔透，与旗袍搭配显得雍容华贵。

An Emerald Necklace Mounted with Diamond

This necklace has green and sparkling beads. Worn by a Chi-pao wearer, it demonstrates her elegant and poised temperament.

- **翡翠镶钻耳环**

此款耳环的材质为翡翠，与旗袍搭配得文雅而端庄，是于简朴中见典雅的一类饰品。

A Pair of Emerald Earrings Mounted with Diamond

This pair of earrings is made of emerald. Belonging to the simple yet graceful type, it can make the wearer of a Chi-pao elegant and dignified.

- **翡翠戒指（20世纪20年代）**

此款铂金镶钻翡翠戒指的一抹水绿，沁人心脾。

An Emerald Ring (1920s)

This is a platinum emerald ring mounted with a diamond. Its greenness gladdens the heart and refreshes the mind of those who see it.

- 镶宝石金簪

簪是中国古代妇女最基本的固定、装饰发型的工具，戴在头上极具古典美。

A Gold Hairpin Mounted with Gem

A hairpin is the most basic hair fastening and decorating tool used by women in ancient China. It gives a classical beauty to its wearer.

- 翡翠镶钻胸花

胸花又称"胸针"，常代替花扣点缀于领口，成为整件旗袍的点睛之笔。胸花大约在清末由国外传入中国，多以黄金、铂金、银、贝壳、宝石、水钻制成，工艺精致。胸花造型较为丰富，多被制成帽子、靴子、衣服、乐器、动植物等造型。

An Emerald Brooch Mounted with Diamond

A brooch is also called breastpin and it is a decoration used at the collarband to replace a flowery button. It can make a Chi-pao even more beautiful. First brought to China by the end of the Qing Dynasty, brooch is mainly made of gold, platinum, silver, shell, gem and rhinestone. There are diversified brooch designs and they are mostly made into the shapes of some commonly seen articles such as hats, boots, clothes, musical instruments, images of animals and plants.

鞋

旗袍能巧妙地修饰女人的身材。比如，旗袍细而窄的腰身使身材显得修长，再配上高跟鞋，能抬高人体的重心，使女人坐、立、行走时自觉地挺胸、收腹，姿态自然就会显得优雅而端庄。女性穿旗袍时适合配高跟皮鞋或布鞋，如果有条件，应该穿真丝绣花或丝绒缎面鞋。鞋的颜色要与旗袍颜色呼应。比如，旗袍上面有金色或银色的亮片，就可以配金色或银色的鞋。

Shoes

Chi-pao can cleverly decorate the different figures of women. For example, the thin waistline of the Chi-pao makes the wearer look slim. A pair of high-heeled shoes can lift the center of gravity of the human body and make the woman stick out her chest and contract her stomach during sitting, standing and walking. In this way, her bearing naturally becomes elegant and decorous. A Chi-pao goes well with high-heeled shoes and cloth shoes. If conditions allow, the real silk embroidered shoes or velvet satin shoes are advisable. The color of the shoes should work in concert with the color of the Chi-pao. For example, if the Chi-pao has golden or silvery sequins, a pair of golden or silvery shoes is the right match.

- **身穿金黄色旗袍的女人**

  在这款礼服旗袍上，纯正的黄色与中国云锦搭配，在款式设计上则采用西式抹胸的款式，中西结合，具有时尚感，同时具有中国特色。金黄色旗袍配以一双具有金属质感的金色高跟鞋，女人的高贵气质尽显无遗。

  ### A Woman in a Golden Yellow Chi-pao
  This formal Chi-pao is made of Chinese brocade in pure yellow. It has a Western undergarment design. The combination of Chinese and Western elements gives the Chi-pao a fashionable look and Chinese characteristics. The woman also wears a pair of golden high-heeled shoes with a metallic appearance, which fully demonstrates her dignified temperament.

绣花鞋

绣花鞋是中国人独创的手工艺品，体现着鞋与刺绣艺术的完美结合。绣花鞋一般以彩色丝线绣成，从鞋头到鞋跟，甚至鞋底和鞋垫，都绣有繁缛、华丽的纹样，图案素材来源于生活，比如花鸟、草虫、飞禽、走兽、山水、风景、戏曲人物等，喻示生活美满。中国古代的汉族女子有缠足的习俗，脚形纤小而屈曲，穿小而尖的绣花鞋，高底一般位于鞋的后部。从20世纪初开始，中国的新女性纷纷不再裹足，多穿平底绣花鞋。绣花鞋的鞋面多以棉布、丝绸等制成，上面的刺绣图案风格多样。

Embroidered Shoes

A pair of embroidered shoes is a handicraft created by the Chinese people. It is the perfect combination of shoes and the art of embroidery. The embroidered shoes are normally made of colorful threads. The entire shoe, from its front to its end and even its sole and pad, is covered with complicated gorgeous patterns. These patterns stem from real life, including the images of flowers, birds, grasses, insects, beasts, landscapes, customs and opera figures, and all represent the happy life. In ancient China, Han women were subject to foot binding. Their feet were made small and curved, suitable for a pair of small and tapered embroidered shoes. A high sole was at the back of the shoe. Since the beginning of the 20th century, women in China no longer bound their feet, but wore flat-sole embroidered shoes. The insteps of the embroidered shoes are mainly made of cloth and silk and embroidered with diversified patterns.

- **清末的平底绣花鞋**

  这两双绣花鞋以锦缎制成，鞋面、鞋帮绣以折枝梅花纹，色彩艳丽，富有传统气息。

  Flat-Sole Embroidered Shoes Made at the End of the Qing Dynasty

  These two pairs of embroidered shoes are made of brocade. Their insteps and uppers are embroidered with the stemmed plum blossom patterns. In bright and diversified colors, they have rich traditional flavors.

扇：旗袍女子与扇子相配构成美丽的风景。中国的扇子一般都是手工制作的，材质主要有竹、木、纸、象牙、翡翠、翎毛等，经过镂雕、烫、钻等手工艺，整个扇面显得非常华美。

Fan: A woman in Chi-pao holding a fan is the most beautiful scene. In China, fans are normally handmade with such materials as bamboo, wood, paper, ivory, emerald and plume. Going through several handicraft processes such as carving, burning, and drilling, the fan will be extremely beautiful.

鞋：白色的高跟鞋带有黑色的细边，鞋的侧面还有三处镂空的装饰，精致而大方，与身上的旗袍搭配得非常合适。

Shoes: Her white high-heeled shoes have black lacing and three hollowed-out decorations on the side. Exquisite and graceful, the shoes are good matches for the Chi-pao.

135

Choosing a Chi-pao Right for You
选购适合自己的旗袍

- 小河边的旗袍女人（20世纪20年代）
A Woman in Chi-pao Standing by a Small River (1920s)

外衣

若在冬天穿旗袍，外衣是必不可少的。如果穿着得当，外衣与旗袍就会相得益彰，显出一种高贵的美感。在冬季穿旗袍比较流行在外面加穿一件长款的风衣或皮草，可选择带有双排扣、蝴蝶结、大翻领等时尚的风衣样式，以很好地体现中西合璧。近几年来，旗袍还有了新的搭配，比如短款旗袍加紧身牛仔裤、旗袍加披肩、中长款旗袍加长款外套、长款旗袍加短外套等。

Overcoat

To wear a Chi-pao in winter, an overcoat is indispensable. Worn properly, the overcoat and the Chi-pao will bring out the best in each other and present a kind of noble beauty. A popular way of wearing a Chi-pao in winter is to put on a long wind coat or a fur coat over the Chi-pao. The wind coat can adopt fashionable styles such as double rows of buttons, bowknot and large lapel, achieving a dressing style combining Chinese and Western fashions. In recent years, new matches for Chi-pao have come into being, including a short Chi-pao with a pair of tight jeans, a Chi-pao with a shawl, a medium and long Chi-pao with a long overcoat, a long Chi-pao with a short overcoat, etc.

- 穿"一口钟"大衣的旗袍女子（20世纪30年代）
  "一口钟"大衣是当时非常流行的旗袍装饰，形似古钟。大衣面料多为丝绸或毛皮，内层絮棉，御寒又美观。

  A Woman Wearing a Chi-pao and a Bell-shaped Overcoat (1930s)
  A bell-shaped overcoat was a very popular decoration for Chi-pao in the 1930s. Made mainly of silk or fur, it has a bat wool lining. It is beautiful and can keep warmth.

手笼：手笼也是与旗袍搭配的小玩意儿，供冬天暖手。富人家用皮草手笼；普通人家则用棉布做的，其上衲有花纹，轻巧又实用。

Muff: Used for keeping hands warm in winter, a muff is a good match for Chi-pao. The rich made their muff with fur material and the poor with cotton. Bearing various patterns, a muff is a light and practical article for women wearing Chi-pao.

- 穿皮草大衣的旗袍女人（20世纪30年代）
  右侧女子怀中的手笼以与大衣相同的皮草制成；右侧的女子身穿褐色貂皮大衣，手戴黑色皮手套，一身暖融融的搭配。

  Two Women Wearing Chi-pao and Fur Coat (1930s)
  The woman on the left holds a muff made of the same fur material as her fur coat. The woman on the right wears a brown marten coat and a pair of black leather gloves, a warm dressing matchup.

穿旗袍应该保持的举止

　　旗袍是一种非常贴合身体的女装，会在一定程度上使穿着者的行动幅度受限。穿上旗袍后，女人的姿态会自动得到修正。比如习惯于弓背的女性会自然挺直身板，配上中高跟鞋，人体重心得以抬高；小腹凸出的女性则会很自觉地收腹……如果穿旗袍时举止不雅，就会失去旗袍的韵味。

　　1.站立时，尽量挺胸、直腰，双手最好自然地合搭在胃的位置或者小腹前。

　　2.坐下时，首先应该撩一撩后裙摆，以免久坐后，旗袍起褶。

　　3.如果坐的是椅子，则只能坐在椅子的前端，不可坐满，坐下后一定不能弓背、塌腰。

- 油画《执扇仕女图》（作者：刘文进）

画上的女人面容清秀，手执扇子，站在芭蕉树下。一袭牙白色薄纱旗袍质地柔软，质朴中透着高洁，展现出了女子的纯洁、娴雅之美。

The Canvas *a Maid Holding a Fan* (by Liu Wenjin)

The woman in the picture has a delicate and pretty look. She holds a fan and stands under a tree. Her white thin gauze Chi-pao has a soft texture and a pure and noble appearance, fully representing the chastity and demure beauty of women.

4.坐下时，双腿不能随意摆放，尽量以优雅的姿势将它们掩藏在裙摆内。如果旗袍的长度刚过膝，那么大腿更要始终闭合。同时还要注意裸露在外的小腿的姿势，最好并拢，向一边倾斜。

5.不论是站着还是坐着，双臂都应该紧贴身体，走路时摆动小臂即可，避免因为幅度过大而露出腋下。

6.如果旗袍的开衩比较高，则需要将步子迈得小一点，以防露出身体。

7.穿旗袍时最好穿连裤丝袜，以防袜口从开衩处显露。

8.有些场合是不适合穿旗袍的，比如穿旗袍骑自行车就非常不雅观。

The Bearing Right for a Chi-pao Wearer

Chi-pao is a women's attire fitting perfectly to the wearer's body. It will to a certain extent limit the wearer's movement. Wearing a Chi-pao, a woman will automatically keep a correct posture. For example, those who have a hunchback will naturally straighten up their back. The medium and high-heeled shoes will lift up the wearer's center of gravity. Those who have a bulged lower abdomen will naturally contract their abdomen. However, if behaving improperly, a Chi-pao wearer won't show the charm of the Chi-pao.

1. When standing, a Chi-pao wearer should stick out her chest and straighten up her waist. Her hands had better naturally hold together and stay at her stomach or lower abdomen.

2. Before sitting down, she should properly arrange her back hem to avoid creases after sitting.

3. If sitting on a chair, she can only

• 穿旗袍时对举止有着严格的要求。
There are strict requirements on behaviors of a Chi-pao wearer.

sit in the front part of the chair. During sitting, she cannot hunch her back or sink her waist.

4. During sitting, her legs cannot be positioned randomly, but should be covered with the hem in a graceful posture. If the Chi-pao just covers her knees, her legs should always stick to each other. Meanwhile, she should keep her exposed shanks close to each other and put them slantly to one side.

5. Either standing or sitting, her two arms should stay close to her body. When walking, she should move her forearms only to avoid exposing her armpits due to large movements.

6. If her Chi-pao has relatively high slits, she'd better take small steps, in case of exposing her body.

7. For a Chi-pao wearer, a panty hose is a good choice to avoid unwanted exposure at the slits.

8. Chi-pao is unsuitable for some activities, such as bicycle riding.

- 2006年4月22日，在北京"2006年驻华使节夫人中国才艺大赛"上，中国专业模特正在进行"旗袍秀"，供参赛的驻华使节夫人们观摩和学习。（图片提供：CFP）

On April 22, 2006, a Chi-pao show was staged at the 2006 Chinese Talent Contest for the Wives of Foreign Diplomatic Envoys in China. Chinese professional models displayed Chi-pao wearing to the contestants.

旗袍的保养

对于买回来的价值不菲的旗袍，妥善的保养、洗涤、收藏都是十分重要的。如今，大多数女性所穿的旗袍都为织锦缎的面料，而这类锦缎都是不宜水洗的。如果在穿着时沾染上了油渍、可乐、口红等，就很难洗掉，因此穿旗袍时要格外注意。

旗袍保养的注意事项

1.不要连续几天穿同一件旗袍，否则容易变形。

2.千万不要将旗袍的袖子卷起来，否则容易变形。

3.穿丝绸制成的旗袍，最好不要贴身，以免被汗液浸蚀，进而变色、变质或破损。

4.穿丝绸制成的旗袍时，要留

Taking Good Care of Your Chi-pao

A Chi-pao is expensive. It is very important to take good care of it, and wash and store it in a proper way. Now, most Chi-pao is made of tapestry satin and such material cannot be washed with water. If stained with oil, coke or lipstick, it is difficult to clean. Therefore, a Chi-pao wearer should pay extra attention to its maintenance.

Tips for Chi-pao Maintenance

1. A Chi-pao cannot be worn for several consecutive days; otherwise, it may deform.

2. It is forbidden to roll up the sleeves of the Chi-pao; otherwise, it may deform.

3. A silk Chi-pao should not be worn next to the skin. This is to avoid sweat erosion, change of color, deterioration, or damage of the Chi-pao.

4. Wearing a silk Chi-pao, one

意尖锐的物件，以免刮伤旗袍，造成钩洞或抽丝。比如，不要穿着旗袍在竹席、藤椅、木板等粗糙的物品上坐或睡觉。

5.由于大多数旗袍是由丝绸制成的，最好干洗，清洗时也一定不要用含酶的洗衣粉。

6.旗袍闲置时，要用宽一点的衣架将之挂起来，注意肩部要撑得妥当。

7.一般来说，将旗袍挂起来比较好。如果没有那么大的空间，则可以把旗袍卷起来，领朝外，这样，它就不会皱，也不会有折线。

8.要在存放旗袍的衣橱里放上防蛀用品。

should stay away from sharp objects to avoid snag or run. For example, she should not sit or lie down on objects with a harsh surface such as bamboo mats, cane chairs and wooden boards.

5. As most Chi-pao is made of silk, dry cleaning should be adopted. In addition, the washing powder containing enzyme should not be used.

6. If not worn, a Chi-pao should be hung up with a relatively wide hanger. The shoulder part should be properly supported.

7. Generally, hanging is a good way to store a Chi-pao. However, if the owner is pressed for storage space, she can roll up her Chi-pao with the collar facing outward. This way, there won't be any creases or fold lines.

8. Some mothproof articles should be used in the wardrobe housing the Chi-pao.

旗袍的清洗

清洗旗袍是一件很复杂的事，不能将它们放到洗衣机里清洗，手洗时也要轻柔。如果有条件，尽量干洗。要特别注意的一点是，有些深色旗袍容易褪色。对于容易褪色的旗袍，我们在清洗之前，可在水中加一点醋或盐，泡一会儿再洗。洗后要把旗袍翻过来晾晒。

Cleaning of a Chi-pao

Cleaning of a Chi-pao requires a complicated procedure. The Chi-pao cannot be washed in a washing machine. It should be handled gently during hand washing. If conditions allow, dry cleaning should be adopted. One should remember that some deep-colored Chi-pao may fade easily. To clean such Chi-pao, add some vinegar or salt to the water and soak the Chi-pao in the water for a while before washing. After washing, the Chi-pao should be turned inside out for airing.

各种旗袍面料的洗涤要点
Tips for Washing the Chi-pao Made of Different Materials

棉布 Cotton Cloth	棉布旗袍可以各种肥皂或洗涤剂洗涤。洗涤前，可将之放在温水中浸泡几分钟，但不宜浸泡得过久。 The Chi-pao made of cotton cloth can be washed with various soups or detergents. Before washing, the Chi-pao can be soaked in warm water for a few minutes, but not too long.
麻布 Linen	麻纤维比较硬，洗涤时用力要小；不可放在洗衣机里绞动，手洗时也不可用力拧干。 Linen has relatively hard fiber and requires gentle washing. The Chi-pao made of linen cannot be washed in a washing machine or wringed too hard with hand.
丝绸 Silk	在清洗丝绸旗袍前，应先将之在温水中浸泡约10分钟，且要选用中性肥皂、中性洗涤剂清洗。洗后轻轻压挤水分，千万不要用力拧。 The silk Chi-pao should be soaked in warm water for about ten minutes before washing. It should be washed with neutral soap flakes or neutral detergents. After washing, water should be gently squeezed out instead of wringing.

雪纺 Chiffon	对于雪纺旗袍，应尽量干洗。如果在宴会上沾上了红酒或油脂，可以用苏打水擦去这些污渍。 The chiffon Chi-pao should be dry-cleaned. If it is stained with red wine or grease, soda water can be used to clean the stains off.
涤纶 Polyester Fiber	在清洗涤纶旗袍前，先要用冷水将之浸泡约15分钟，然后再用一般合成洗涤剂洗涤。清洗完，应将之放置在阴凉、通风处晾干，不可暴晒，不宜烘干。 The Chi-pao made of polyester fiber should be soaked in cold water for about fifteen minutes before being washed with ordinary synthetic detergents. The washed Chi-pao should be placed in a cool and well-ventilated place for airing. It should not be put under intense sunshine or be dried beside or over a fire.
毛涤混纺 Wool / Polyester Blend Fabric	在洗涤毛涤混纺旗袍前，先用冷水将之浸泡2—3分钟，然后再对重点部位进行揉洗。洗后将之弄平整，挂在通风处阴干。 Before washing, the Chi-pao made of wool / polyester blend fabric should be soaked in cold water for two to three minutes. Then, some key areas shall be rubbed with hands. After washing, the Chi-pao should be leveled and hung in a well-ventilated place for airing.
刺绣 Embroidery	清洗有刺绣的旗袍时一定格外小心；在旗袍晾干后，最好用一块布垫在刺绣图案上，然后用熨斗整烫一遍。 Special care should be taken when washing a Chi-pao with embroidery. After washing and drying, a piece of cloth had better be placed on the embroidery pattern before ironing the entire Chi-pao.
手绘 Hand Painted Fabric	清洗手绘旗袍时不要使用漂白性的洗涤剂，不要用力搓图案的表面，最好用40°C以下的温水或凉水清洗。 When washing a Chi-pao made of hand painted fabric, no bleaching detergent should be used. Neither should one rub hard the surface of the pattern. It is better to wash the Chi-pao with warm water under 40°C (104 degrees Fahrenheit) or cool water.

附录：现代旗袍鉴赏
Appendix: Appreciation of Modern Chi-pao

- **红色改良旗袍**

这是一款改良的旗袍礼服，云锦配雪纺，面料的软硬搭配适宜，整体效果雍容而大方，又不失青春气息。

A Red Modified Chi-pao

This modified formal Chi-pao is made of brocade and chiffon. The proper use of both soft and hard materials makes the Chi-pao tasteful and youthful.

• **深蓝色无袖传统旗袍**
这是一款丝绒面料的传统旗袍，采用深蓝色的面料，体现了现代知识女性的优雅。
A Dark Blue Sleeveless Traditional Chi-pao
This is a traditional Chi-pao made of dark blue velvet. It displays the elegance of modern intellectual women.

- **红色大泡泡袖旗袍**

红色大泡泡袖搭配旗袍是比较少见的。这款旗袍以红色织锦缎制成，蟠龙云锦与海水图案搭配，给人以庄重、大方之感。同时，泡泡袖的设计增添了旗袍的时尚感。蟠龙和海水是戏曲服装的图案，因而它可以作为旗袍礼服来穿着。可以说，这件旗袍礼服很好地体现了古典与现代的完美结合。

A Red Chi-pao with Large Puffed Sleeves

It is rare for red large puffed sleeves to be used on a Chi-pao. This Chi-pao is made of red tapestry satin and brocade with loong and sea water patterns, which give the viewers a solemn and graceful feeling. The puffed sleeves add fashionable sense to the Chi-pao while the loong and sea water patterns, which are commonly used on opera costumes, make it a formal attire. It is safe to say that this Chi-pao is a perfect combination of classics and modernity.

- **黑色蟠龙云纹旗袍**

这款旗袍采用黑色的缎面与黑色蟠龙云纹搭配的形式，在搭配方法上就给人以美的感受。同时，两种面料的搭配使旗袍显得庄重、大方，黑色蟠龙云纹上又被做了珠绣，增加了华丽感和工艺美。

A Black Chi-pao with Loong and Cloud Patterns

This Chi-pao is made of black satin with black loong and cloud patterns, a mixture guaranteeing the feeling of beauty. Meanwhile, the two materials are used together to make the Chi-pao solemn and tasteful. The black loong brocade bears bead embroidery, which adds a gorgeous sense and an artistic beauty to the Chi-pao.

• 宝蓝色旗袍礼服

这款旗袍以宝蓝色锦缎作为面料，上身装饰有如意云纹，使整件旗袍显得非常华丽、高贵。

A Sapphire Blue Formal Chi-pao

This Chi-pao is made of sapphire blue brocade. Its upper part has auspicious cloud patterns, which make the entire Chi-pao gorgeous and valuable.

● **紫色丝绒旗袍**
这款旗袍紫色的丝绒面料上被缝制了金色的珠绣，体现了女性的雍容华贵、典雅和大方。
A Purple Velvet Chi-pao
This Chi-pao is made of purple velvet with golden bead embroidery. Such a design demonstrates women's elegant, poised and graceful bearing.

• 深红色长旗袍

这款旗袍腰部装饰有如意云纹，为纯手工制作，还点缀以珠绣，有着很强的传统特色，给人以高贵、典雅之感。

A Deep Red Long Chi-pao

This Chi-pao has auspicious cloud patterns on its waist. Made purely by hand and with bead embroidery decoration, it shows strong traditional characteristics and has around it a noble and elegant atmosphere.

- **玫红色传统旗袍**

  这是一款玫红色的传统旗袍，采用双滚边设计，弧形造型的前襟强调女性的柔和曲线，色彩搭配也很和谐。

  A Rose Traditional Chi-pao

  This is a rose traditional Chi-pao with double-lacing design. Its arc-shaped front piece emphasizes the gentle curve of women. In addition, this Chi-pao has harmonious color application.

- **黑色改良旗袍**

  在这款黑色的改良旗袍中，设计者结合立体裁剪的方法，设计出了两件套的感觉，体现了现代女性的干练。

  A Black Modified Chi-pao

  This black modified Chi-pao is made with 3D tailoring method. It looks like a two-piece dress, suitable for modern women who have both capability and experience.

• **黑色抹胸旗袍**

这款旗袍采用手绣牡丹搭配西式抹胸的造型,给人以经典、华丽之感。

A Black Undergarment-style Chi-pao

This Chi-pao has the Western undergarment design and is decorated with a hand-embroidered peony pattern. It looks classical and gorgeous.

- **红色锦缎旗袍**

  这是一款红色锦缎配云锦纹抹胸旗袍礼服，中式面料与西式款型结合，是晚会着装的不错选择。

  A Red Brocade Chi-pao

  This is a red undergarment-style formal Chi-pao made of brocade. The combination of Chinese material and Western design makes it a good choice for evening parties.

• 黑色改良旗袍

这款旗袍以黑色的织锦缎配云纹，同时加入了雪纺面料。多种面料的应用打破了旗袍面料的单一性，柔与刚的结合、中西文化的互搭，使这款旗袍成为知识女性在参加社交活动时的最佳选择。

A Black Modified Chi-pao

This Chi-pao is made of black tapestry satin with cloud patterns. It also contains chiffon. The application of several materials breaks the unitary sense given by unitary material. The combination of softness and firmness and the merging of Chinese and Western cultures make this Chi-pao the ideal choice for intellectual women during social activities.

红色手绣改良旗袍

这款旗袍的手绣图案部分由海水、牡丹、凤凰这些具有美好寓意的纹样组成,喻示爱情像海水一样绵绵不绝,生活繁花似锦、富贵、平安。前襟的弧形蓝色镶边全部由手工制作完成,制作工艺非常精湛。

A Red Modified Chi-pao with Hand Embroidered Patterns

The hand embroidered patterns on this Chi-pao are all auspicious images such as sea water, peony and phoenix. They symbolize the love that goes on endlessly like the sea water and the life that is prosperous, wealthy, and peaceful. The arc blue lace on the front piece is completely handmade, showing extremely exquisite workmanship.

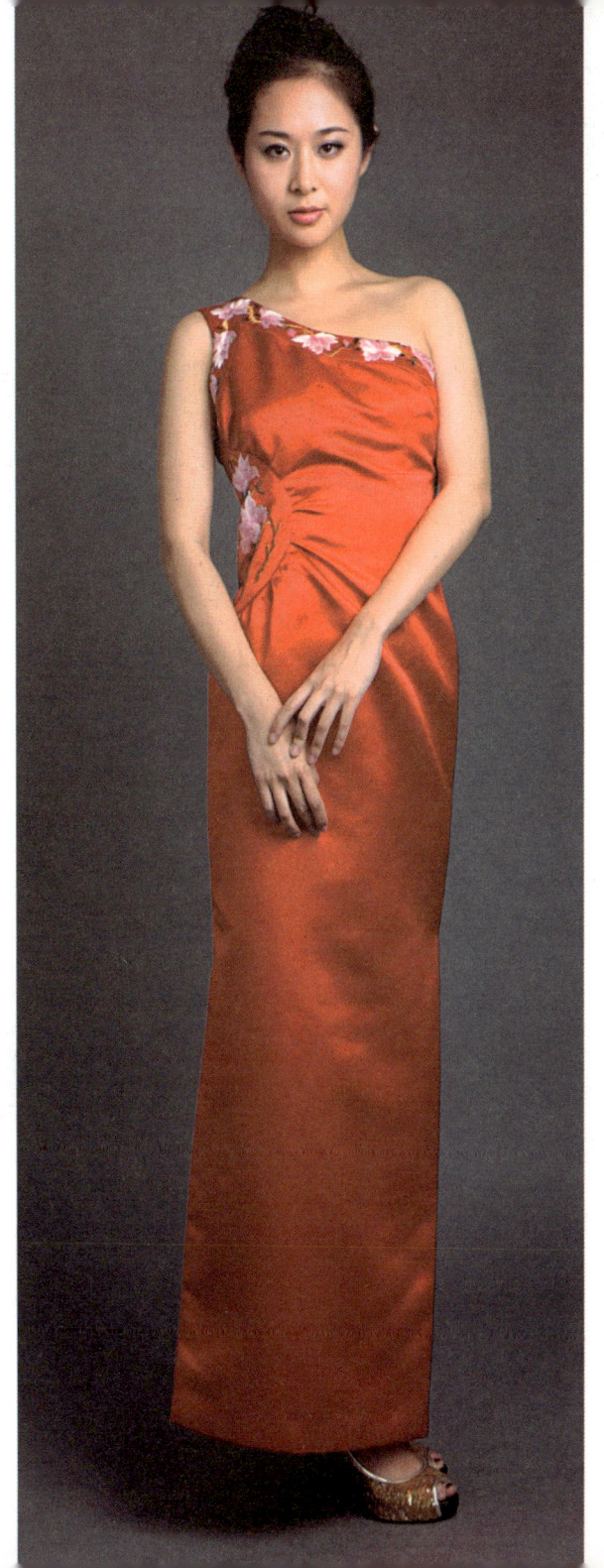

附录：现代旗袍鉴赏 Appendix: Appreciation of Modern Chi-pao

- **红色单肩旗袍**

  这是一款单肩手绣旗袍，手绣的玉兰花活灵活现，单肩的设计突出了高贵和典雅。穿着这件旗袍礼服，自然会成为宴会的焦点。

  A Red One-shoulder Chi-pao

  This is a one-shoulder Chi-pao with vivid hand embroidered *yulan* patterns. The one-shoulder design gives a noble and elegant atmosphere, making the wearer of this Chi-pao the focus of any banquet.

- **白色手绣旗袍**

  这款旗袍的腰部手绣图案汲取了戏曲服装的一些元素，白色的面料犹如一张白纸，荷花跃然其上，给人以清秀之感。

  A White Chi-pao with Hand Embroidered Patterns

  The hand embroidered patterns on the waist of this Chi-pao absorb some elements from opera costumes. The white material is like a white piece of paper. The lotus flower on it looks fresh and graceful.

- **黑色短旗袍**

  这款旗袍腰部绘有缠枝藤，使一件普通的黑色旗袍礼服具有了鲜明的个性。

  A Black Short Chi-pao

  This Chi-pao has hand painted branch-winding vines on its waist. The patterns give this ordinary black Chi-pao distinct characteristics.

- 灰色丝绒旗袍

这款旗袍采用灰色丝绒与白色蕾丝花边搭配的形式，细节精美，典雅而高贵。

A Grey Velvet Chi-pao

This Chi-pao is made of grey velvet with white lace. It shows a strong sense of beauty in details, looking elegant and noble.

- 红色改良旗袍

这款旗袍织锦缎与蕾丝的搭配彰显着中西文化的结合，对红色部分采取褶皱的设计，配合黑色蕾丝，给人以神秘、优雅之感。

A Red Modified Chi-pao

This Chi-pao is made of tapestry satin with lace, which showcases the combination of Chinese and Western cultures. The red part adopts the fold design accompanied with black lace. Such arrangement guarantees a mysterious and graceful feeling.

附录：现代旗袍鉴赏 Appendix: Appreciation of Modern Chi-pao

● 灰色蕾丝旗袍

这款旗袍采用仿蕾丝棉布与蕾丝披肩搭配的形式，样式新颖，穿起来更显身姿曼妙，是日常穿着的极佳选择。

A Grey Lace Chi-pao

Made of lace-like cotton cloth, this Chi-pao shows a novel design. Accompanied with a lace shawl, it can better display the fine figure of its wearer. It is an ideal dress for everyday wearing.

● 玫红色传统旗袍

这款旗袍采用无袖设计，配以传统图案，与身形贴合得十分完美，显得经典而大方。

A Rose Traditional Chi-pao

This Chi-pao has a sleeveless design and bears some traditional patterns. Fitting the body of its wearer perfectly, it is indeed a classical and graceful dress.